J. J. van Rensburg.

Sept 1978

AF698496

LIFE AND OTHER CONTINGENCIES

This volume forms part of a series of text-books published under the authority of the Institute of Actuaries and the Faculty of Actuaries and is designed to meet the needs of students preparing for the actuarial examinations.

LIFE AND OTHER CONTINGENCIES

BY

P. F. HOOKER
F.I.A.

AND

L. H. LONGLEY-COOK
M.A., F.I.A., F.C.A.S., A.S.A.

VOLUME II

CAMBRIDGE

Published for the Institute of Actuaries and the Faculty of Actuaries

AT THE UNIVERSITY PRESS

1957

Published by the Syndics of the Cambridge University Press
Bentley House, 200 Euston Road, London NW1 2DB
American Branch: 32 East 57th Street, New York, N.Y.10022

ISBN: 0 521 05327 7

First published 1957
Reprinted 1973

First printed in Great Britain at the University Press, Cambridge
Reprinted in Great Britain by
Redwood Press Limited, Trowbridge, Wiltshire

CONTENTS

ABBREVIATIONS

J.I.A.	*Journal of the Institute of Actuaries*
J.S.S.	*Journal of the Institute of Actuaries Students' Society*
T.A.S.A.	*Transactions of the Actuarial Society of America*
T.S.A.	*Transactions of the Society of Actuaries*
Proc. Cent. Assembly	*Proceedings of the Centenary Assembly of the Institute of Actuaries*

PREFACE

At the time when Spurgeon's *Life Contingencies* was first published, the study of multiple-decrement tables was very much in its infancy; it is not surprising therefore that Spurgeon disposed of the subject in one short chapter. In the present volume the first two chapters (numbered 19 and 20 because Volume I contained 18 chapters) are devoted to the subject. Chapter 19 deals with the multiple-decrement table by itself and Chapter 20 with the relationships between the multiple-decrement table and its 'family' of single-decrement tables. The authors are greatly indebted to Messrs Bailey and Haycocks, authors of *Some theoretical aspects of Multiple Decrement Tables*, for many of the ideas underlying Chapter 20. They have, however, sought to develop the subject in their own way; in particular they have felt it important to establish an analogy between the model and actual experience and to make use of this analogy to demonstrate the identity of the forces of decrement.

In Chapter 21 the student is introduced to the joint-life mortality table; some standard deviations which are of interest but hardly merit a chapter to themselves are included.

Chapters 22 (joint-life and last survivor annuities, assurances, etc.), 24 (contingent assurances) and 25 (reversionary annuities) follow conventional lines and contain little that is new; the authors have purposely given less emphasis to complicated functions than was at one time the fashion. In Chapter 23 they have developed some general theorems of which many of the formulae in the book are special cases.

Spurgeon, in his *Life Contingencies*, gave the Griffith Davies definition of D_{xy}:

$$D_{xy} = v^x l_x l_y$$

as well as the De Morgan definition:

$$D_{xy} = v^{\frac{x+y}{2}} l_x l_y.$$

Only the latter definition is recognized in the International Notation but in practice most compilers of tables have preferred the former definition in spite of the objections which Spurgeon mentions. The authors have followed Spurgeon's example by giving both definitions in Chapter 26, in which commutation functions and approximate integration are discussed, and they have suggested a simple convention for removing the objections. Although they have tried to remain loyal to the International Notation, they cannot conceal their hope that one day the Griffith Davies definition, with the convention which they suggest, may be officially adopted.

Various transformations and other methods of evaluation, including those which depend on a 'law' of mortality, are described in Chapter 27.

In Chapters 28 and 29 the authors have developed, by modern methods, formulae on the lines of those which were first set out by George King in *J.I.A.* **39** and have been used in pension fund valuations ever since. They hope that the new approach, although leading to the same results as the old-fashioned method, will give students a firmer foundation on which to base their work. Chapters 30 and 31 deal briefly with widows' and orphans' funds and disability benefits respectively.

As in Volume I, a few examples have been appended to the chapters. These are not intended to take the place of a tuition course but merely to help the student to assimilate what he has read in the chapter and to prepare him for the more detailed study, under the supervision of a tutor, which is necessary to bring him up to examination standard. The student is strongly urged to work out all these examples for himself, *including the arithmetic*, before reading the printed solutions. It cannot be too strongly emphasized that a theoretical knowledge of the subject, however thorough, is valueless if it is accompanied by an inability to recognize a gross inaccuracy in an arithmetical result. For this reason a rough check should always be applied; the fact that rough checks have not been appended to all the solutions printed in this book does not mean that the authors consider them unnecessary.

A few sections have been marked with asterisks to indicate that the authors, while advising the student to read these sections, do not regard them as coming within the syllabus of those portions of the examinations which the book is principally intended to cover.

The authors again wish to acknowledge the great debt which they owe to George King, E. F. Spurgeon and many other writers on whose works they have drawn freely. They would also like to express their appreciation of the help which they have received from some of their colleagues and especially from Messrs J. L. Anderson and H. Hosking Tayler. A special word of thanks is due to Mr B. P. Willcox, who has read the proofs of both Volumes and detected several blunders which might otherwise have escaped notice until after publication.

P. F. H., L. H. L-C.

March 1957

CHAPTER 19

THE MULTIPLE-DECREMENT TABLE

19·1. Introduction

The theory of Life Contingencies developed in Volume I of this book was based on a mortality table in which the only mode of decrement was death. When assumptions as to other contingencies, such as marriage and sickness, were introduced, these assumptions merely took the form of proportions which were applied to the functions of the mortality table; the values of l_x were unaffected and corresponded to the empirical sequence n_x described in Chapter 1, i.e. the number of persons assumed to survive to successive ages out of a group of persons kept continuously under observation until death.

In many actuarial problems we are concerned not with the survival (in the normal sense of this word) of a life or group of lives but with survival as a member or members of a designated class. Such a class may be the residents of a country where the population will be subject to depletion by emigration as well as by death; or policyholders in a life office where termination may occur as a result of lapse or surrender as well as death; or employees of a company who are members of a pension scheme and who may cease to be employees as a result of withdrawal or retirement. It follows that, although (as we have seen in Volume I) the single-decrement mortality table serves as a useful model for a great deal of actuarial work, a table with two or more modes of decrement would correspond more closely with actual experience. Such a table is known as a multiple-decrement table; if there are only two modes of decrement it is often referred to as a double-decrement table.

For reasons which need not concern us here it is not usual to employ multiple-decrement tables in connexion with population problems or in ordinary life office work. The main use of such tables is in connexion with schemes which provide benefits on retirement, marriage, disablement, etc. However, in the present chapter we shall be concerned with the theory of the tables rather than with the uses to which they are put.

Just as in Volume I we set up a mathematical model with the fundamental function l_x corresponding to the empirical sequence n_x, the survivors to age x out of a larger number of persons n_α, so we can set up a mathematical model with the fundamental function $(al)_x$ corresponding to the empirical sequence $(an)_x$, the survivors to age x, as members of a designated class, out of a larger number of persons $(an)_\alpha$ who were members of the class at age α.

It is possible to set up a mathematical model to represent the experience of inanimate objects such as motor-cars or aeroplanes which are subject to decrement by reason of accident, wearing out of irreplaceable parts, obsolescence, etc. However in this book we shall be concerned only with the survival of human beings as members of a class.

19·2. Notation

There is no recognized International Notation for multiple-decrement tables. The notation used in this chapter will be that which has been developed in Great Britain during recent years. The characteristic feature of this notation is that a letter (in most of this chapter the letter *a*) is prefixed to each symbol and that this letter and the principal letter of the symbol to which it is prefixed are enclosed in brackets. For example, the symbol $(al)_x$ represents a hypothetical function which is like the l_x of a single-decrement table but is subject to more than one decrement, and $(ad)_x$ represents $-\Delta(al)_x$ just as d_x represents $-\Delta l_x$.

A symbol which relates to only one mode of decrement has a letter attached to the upper right-hand corner to indicate the mode of decrement. In practical work, *d* is used for death, *w* for withdrawal, *i* for ill-health retirement, and so on; thus $(ad)_x^d$ denotes the portion of $(ad)_x$ which is attributed to the mode of decrement death. In theoretical work, the Greek letters α, β, γ, ... are used to indicate different modes of decrement whose nature is not specified; thus $(ad)_x^\alpha$ denotes the portion of $(ad)_x$ which is attributed to the mode of decrement α. It follows that

$$(ad)_x = (al)_x - (al)_{x+1} \qquad (19\cdot1)$$

and that $(ad)_x = (ad)_x^\alpha + (ad)_x^\beta + (ad)_x^\gamma + \ldots.$

The summation of $(ad)_x^\alpha$ over all modes of exit will be represented by $\sum_{\kappa=\alpha\ldots} (ad)_x^\kappa$ so that the latter relationship may be written

$$(ad)_x = \sum_{\kappa=\alpha\ldots} (ad)_x^\kappa. \qquad (19{\cdot}2)$$

It should be noted that the letter α is used in two different senses according to the position which it occupies in relation to the main symbol. When it is attached to the upper right-hand corner of the symbol it indicates a mode of decrement; when it is attached to the lower right-hand corner it stands for the youngest age in the table.

19·3. An example of a multiple-decrement table

The following extract from a multiple-decrement table will serve as an illustration. It is taken from Section VIII of *Actuarial Tables for Examination Purposes*; the headings of the columns (which, in the original table, are in an abbreviated notation commonly used in pension fund work) have been altered to bring them into line with the notation of this chapter.

TABLE 19·1

Age x	$(al)_x$	$(ad)_x^w$	$(ad)_x^d$
20	100,000	2,998	155
21	96,847	2,613	155
22	94,079	2,256	155
23	91,668	1,923	156
24	89,589	1,611	157
25	87,821	1,316	159
...	...	...	...

It will be seen that, just as l_{x+1} in a single-decrement table may be obtained by deducting d_x from l_x, so $(al)_{x+1}$ in this double-decrement table may be obtained by deducting $(ad)_x^w$ and $(ad)_x^d$ from $(al)_x$.

The figures in the column headed $(al)_x$ in the foregoing table may be regarded as representing the numbers of lives who are expected to remain members of the class (in this case employees covered by a pension fund) at successive integral ages out of 100,000 who were members of the class at age 20; similarly the figures in the columns headed $(ad)_x^w$ and $(ad)_x^d$ may be regarded as representing the numbers of lives, out of the 100,000, who are expected to cease to be members of the class in successive years of age by reason of withdrawal and death respectively.

19·4. Forces of decrement

Although the figures in the multiple-decrement table may be regarded as representing numbers of lives, it is important not to lose sight of the fact that the multiple-decrement table, like the single-decrement table of Volume I, is a mathematical model. $(al)_x$ is a purely hypothetical function which is assumed to decrease as x increases.

The ratio which the rate of decrease of $(al)_x$ at any age x bears to the value of $(al)_x$ at that age is denoted by the symbol $(a\mu)_x$ and may be referred to as the total force of decrement at age x.

$$(a\mu)_x = -\frac{1}{(al)_x}\frac{d(al)_x}{dx} \tag{19·3}$$

$$= -\frac{d}{dx}\log_e(al)_x. \tag{19·4}$$

Let $(al)_x^\alpha$, $(al)_x^\beta$, ... be defined by the statement that out of $(al)_x$ lives who are members of the (al) class at age x it is expected that $(al)_x^\alpha$ will ultimately leave that class by mode of exit α, $(al)_x^\beta$ will ultimately leave by mode β, and so on. It follows that

$$\left.\begin{aligned}(al)_x^\alpha &= \sum_{t=0}^{\infty}(ad)_{x+t}^\alpha,\\ (al)_x^\beta &= \sum_{t=0}^{\infty}(ad)_{x+t}^\beta,\end{aligned}\right\} \tag{19·5}$$

and so on, and that
$$(al)_x = \sum_{\kappa=\alpha\ldots}(al)_x^\kappa. \tag{19·6}$$

The ratio which the rate of decrease of $(al)_x^\alpha$ at any age x bears to the value of $(al)_x$ at that age is denoted by the symbol $(a\mu)_x^\alpha$ and is referred to as the force of decrement by mode α at age x.

$$(a\mu)_x^\alpha = -\frac{1}{(al)_x}\frac{d(al)_x^\alpha}{dx}. \qquad (19\cdot7)$$

$(a\mu)_x^\beta$, $(a\mu)_x^\gamma$, ... are defined similarly.

Differentiating both sides of formula (19·6) with respect to x, dividing throughout by $(al)_x$ and changing the sign, we have

$$(a\mu)_x = \sum_{\kappa=\alpha\ldots} (a\mu)_x^\kappa. \qquad (19\cdot8)$$

In practical work, when the nature of the decrements is specified as death, withdrawal, etc., instead of being denoted by the arbitrary symbols α, β, γ, ..., $(a\mu)_x^d$ is referred to as the force of mortality, $(a\mu)_x^w$ as the force of withdrawal, and so on.

Little use is found for the total force of decrement $(a\mu)_x$ but, as will be seen in the next chapter, the partial forces of decrement $(a\mu)_x^\alpha$, $(a\mu)_x^\beta$, ... are of considerable importance. The functions $(al)_x^\alpha$, $(al)_x^\beta$, ... rarely occur in practice and have been introduced in this section for the purpose of defining the partial forces of decrement.

In Volume I it was mentioned that for theoretical work it was convenient to assume that l_x was a continuous function having unique finite differential coefficients to any required order at all points. In a multiple-decrement table it is not always possible to make this assumption. For instance, some pension schemes provide that members may retire immediately on attaining age 60, or at any time thereafter, regardless of their state of health, but that they may not retire before attaining age 60 unless they have suffered a breakdown of health. In a table designed to represent the experience of such a class of persons, the assumption that $d(al)_x/dx$ is unique and finite at age 60 would usually be inappropriate. The diagram shown on page 6 illustrates this point.

In these circumstances $(al)_x$ is not fully defined in terms of the forces of decrement and the radix; it is necessary to know also the proportion of persons who retire at exact age 60. This point is of no practical importance but it occasionally introduces a complication in the theoretical development of the subject.

19·5. Relationships between total decrement functions

It will be seen that the relationships between the functions $(al)_x$, $(ad)_x$ and $(a\mu)_x$ of the multiple-decrement table are exactly the same as those between the functions l_x, d_x and μ_x of the mortality table. So long as we are concerned with total decrements only, the

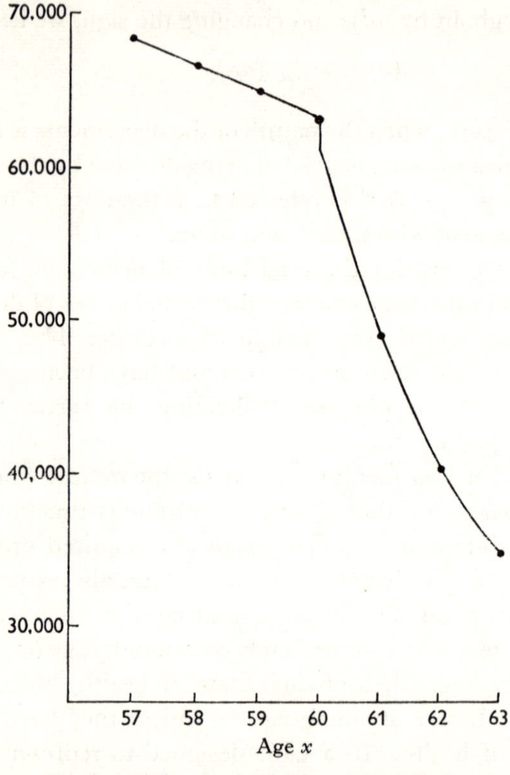

FIG. 19·1. Portion of the graph of $(al)_x$—Multiple-decrement Table in *Actuarial Tables for Examination Purposes*

(The table contains values for integral ages only; the drawing of the graph between $(al)_{60}$ and $(al)_{61}$ is arbitrary.)

developments of the mortality table which were set out in Volume I of this book will apply equally to the multiple-decrement table. In particular ${}_n(ap)_x$ denotes the probability that a life aged x who is a member of the (al) class will still be a member of that class in

n years' time, and ${}_n(aq)_x$ denotes the probability that a life aged x who is a member of the (al) class will leave by any mode of decrement between age x and age $x+n$. As in the mortality table, the prefix is dropped from these symbols when $n=1$.

It will not be necessary to repeat the development of Volume I but it may be helpful to set out some of the more important results:

$$(al)_x=\int_0^\infty (al)_{x+t}(a\mu)_{x+t}\,dt, \tag{19·9}$$

$${}_n(ap)_x=\frac{(al)_{x+n}}{(al)_x}, \tag{19·10}$$

$$(ap)_x=\frac{(al)_{x+1}}{(al)_x}, \tag{19·11}$$

$${}_n(ap)_x=\exp\left\{-\int_0^n (a\mu)_{x+t}\,dt\right\}, \tag{19·12}$$

$$\left.\begin{aligned}{}_n(aq)_x&=1-{}_n(ap)_x\\&=\frac{(al)_x-(al)_{x+n}}{(al)_x},\end{aligned}\right\} \tag{19·13}$$

$$(aq)_x=1-(ap)_x=\frac{(ad)_x}{(al)_x}, \tag{19·14}$$

$$(ad)_x=\int_0^1 (al)_{x+t}(a\mu)_{x+t}\,dt, \tag{19·15}$$

$$(aq)_x=\int_0^1 {}_t(ap)_x(a\mu)_{x+t}\,dt. \tag{19·16}$$

The total central rate of decrement at age x is

$$(am)_x=\frac{\int_0^1 (al)_{x+t}(a\mu)_{x+t}\,dt}{\int_0^1 (al)_{x+t}\,dt}. \tag{19·17}$$

Writing
$$(aL)_x=\int_0^1 (al)_{x+t}\,dt, \tag{19·18}$$

we have
$$(am)_x=\frac{(ad)_x}{(aL)_x}. \tag{19·19}$$

19·6. Relationships between single-decrement functions

In the previous section we considered the relationships between total-decrement functions such as $(al)_x$ and ${}_n(ap)_x$. We have now to develop the relationships which apply to the single-decrement functions $(ad)_x^\alpha$, ${}_n(aq)_x^\alpha$, etc. of the multiple-decrement table.

It was shown in section 2 that

$$(ad)_x = \sum_{\kappa=\alpha\ldots} (ad)_x^\kappa$$

and in section 4 that $\quad (a\mu)_x = \sum_{\kappa=\alpha\ldots} (a\mu)_x^\kappa.$

From formula (19·7) we have

$$(al)_x^\alpha = \int_0^\infty (al)_{x+t}(a\mu)_{x+t}^\alpha \, dt. \tag{19·20}$$

The symbol ${}_n(ap)_x^\alpha$ is not used but ${}_n(aq)_x^\alpha$ denotes the probability that a life aged x who is a member of the (al) class will leave by mode of decrement α between ages x and $x+n$. It follows that

$${}_n(aq)_x^\alpha = \frac{(al)_x^\alpha - (al)_{x+n}^\alpha}{(al)_x} \tag{19·21}$$

and that

$${}_n(aq)_x = \sum_{\kappa=\alpha\ldots} {}_n(aq)_x^\kappa. \tag{19·22}$$

When $n=1$ the prefix is dropped and we write

$$(aq)_x^\alpha = \frac{(ad)_x^\alpha}{(al)_x} \tag{19·23}$$

and

$$(aq)_x = \sum_{\kappa=\alpha\ldots} (aq)_x^\kappa. \tag{19·24}$$

From formula (19·20)

$$(ad)_x^\alpha = \int_0^1 (al)_{x+t}(a\mu)_{x+t}^\alpha \, dt, \tag{19·25}$$

whence

$$(aq)_x^\alpha = \int_0^1 {}_t(ap)_x (a\mu)_{x+t}^\alpha \, dt. \tag{19·26}$$

Similarly

$${}_n(aq)_x^\alpha = \int_0^n {}_t(ap)_x (a\mu)_{x+t}^\alpha \, dt. \tag{19·27}$$

The central rate of decrement by mode of exit α is defined by the relationship

$$(am)_x^\alpha = \frac{\int_0^1 (al)_{x+t}(a\mu)_{x+t}^\alpha \, dt}{\int_0^1 (al)_{x+t} \, dt} = \frac{(ad)_x^\alpha}{(aL)_x}. \tag{19·28}$$

It follows that

$$(am)_x = \sum_{\kappa=\alpha\ldots} (am)_x^\kappa. \tag{19·29}$$

19·7. Approximate expressions for central rates of decrement

The importance of central rates of decrement lies in their similarity to the form in which measures of mortality, withdrawal and other decrements are often obtained in experience investigations. It is frequently convenient to be able to express the central rates $(am)_x$, $(am)_x^\alpha$, etc., in terms of the probabilities $(aq)_x$, $(aq)_x^\alpha$, etc., and of the forces $(a\mu)_x$, $(a\mu)_x^\alpha$, etc., and vice versa. The necessary relationships will be established in this section. For convenience the relationships between the total-decrement functions $(am)_x$, $(aq)_x$ and $(a\mu)_x$ will be deduced here although they correspond to those which have already been demonstrated in Volume I in respect of the mortality table.

On the assumption that the total decrements are distributed uniformly over the year of age, we have from formula (19·28)

$$\left.\begin{aligned} (am)_x^\alpha &\doteq \frac{(ad)_x^\alpha}{(al)_x - \frac{1}{2}(ad)_x}, \\ \text{whence} \qquad (am)_x^\alpha &\doteq \frac{2(aq)_x^\alpha}{2-(aq)_x}. \end{aligned}\right\} \qquad (19·30)$$

From formula (19·30), summing for all modes of exit, or from formula (19·19), we have

$$(am)_x \doteq \frac{2(aq)_x}{2-(aq)_x}, \qquad (19·31)$$

whence

$$(aq)_x \doteq \frac{2(am)_x}{2+(am)_x}. \qquad (19·32)$$

From formula (19·30)

$$(aq)_x^\alpha \doteq (am)_x^\alpha\{1-\tfrac{1}{2}(aq)_x\},$$

whence, by substituting for $(aq)_x$ the expression on the right-hand side of formula (19·32), we have

$$(aq)_x^\alpha \doteq \frac{2(am)_x^\alpha}{2+(am)_x}. \qquad (19·33)$$

On the same assumption that the total decrements are distributed uniformly over the year of age, the denominator in formula (19·17) becomes equal to $(al)_{x+\frac{1}{2}}$; the product $(al)_{x+t}(a\mu)_{x+t}$ in the numerator becomes constant for all values of t between 0 and 1 and therefore the numerator can be replaced by $(al)_{x+\frac{1}{2}}(a\mu)_{x+\frac{1}{2}}$. Hence we have the approximation

$$(am)_x \doteq (a\mu)_{x+\frac{1}{2}}. \tag{19·34}$$

If we further assume that the decrements by mode α also are distributed uniformly over the year of age we have from formula (19·28)

$$(am)_x^{\alpha} \doteq (a\mu)_{x+\frac{1}{2}}^{\alpha}. \tag{19·35}$$

This approximate relationship is of considerable practical importance.

19·8. Monetary functions

Commutation functions can be constructed in connexion with multiple-decrement tables. For example, if we write

$$(aD)_x = v^x (al)_x$$

and

$$(aN)_x = \sum_{t=0}^{\infty} (aD)_{x+t},$$

the value of an annuity-due to (x), payable so long as he remains a member of the (al) class, is

$$\frac{(aN)_x}{(aD)_x}.$$

Also, if we write

$$(aC)_x^d = v^{x+1}(ad)_x^d$$

and

$$(aM)_x^d = \sum_{t=0}^{\infty} (aC)_{x+t}^d,$$

the value of an assurance of 1, payable at the end of the year of death of (x) if he dies while in the (al) class, is

$$\frac{(aM)_x^d}{(aD)_x}.$$

The main purpose for which commutation functions based on multiple-decrement tables are required is the valuation of pension funds and an elaborate technique for this purpose has been evolved. Full consideration of monetary functions will therefore be deferred to Chapters 28 and 29.

19·9. Combined tables

Occasionally we require a model which may be used to represent not only the experience of persons so long as they remain members of a certain class but also their continued experience as members of some related class. The most usual tables of this type are a combined mortality and marriage table and a combined mortality and disability table. The following is an extract from a combined mortality and marriage table.

TABLE 19·2

Age	Number expected to attain age x		Number of decrements between ages x and $x+1$		
	Bachelors $(bl)_x$	Married men $(hl)_x$	Bachelors dying $(bd)_x^d$	Bachelors marrying $(bd)_x^h$	Married men dying $(hd)_x^d$
...	...	...	...	...	...
30	5,367	3,771	42	393	25
31	4,932	4,139	43	361	27
32	4,528	4,473	46	330	29
...	...	...	...	...	...

The term 'married' includes widowers and divorced men; the letter h stands for 'husbands'.

It will be seen that the columns headed $(bl)_x$, $(bd)_x^d$ and $(bd)_x^h$ form a double-decrement table, the class being bachelors and the decrements being death and marriage. The columns headed $(hl)_x$, $(bd)_x^h$ and $(hd)_x^d$ form another table in respect of the married; this table is like a double-decrement table but has one increment, bachelors marrying, and one decrement, married men dying. The formulae which have been deduced in this chapter can therefore be applied to the bachelors' portion of the mortality and marriage table; they can also be applied to the married men's portion if $(bd)_x^h$ is treated as a 'negative decrement'.

Thus, from formula (19·7), the force of marriage for bachelors is

$$\left.\begin{aligned}(b\mu)_x^h &= -\frac{1}{(bl)_x}\frac{d}{dx}\left\{\sum_{t=0}^{\infty}(bd)_{x+t}^h\right\}\\ &\doteq \frac{(bd)_{x-1}^h+(bd)_x^h}{2(bl)_x}.\end{aligned}\right\} \qquad (19{\cdot}36)$$

From formula (19·23), the probability that a bachelor will marry within a year is

$$(bq)_x^h = \frac{(bd)_x^h}{(bl)_x} \qquad (19{\cdot}37)$$

and, from formula (19·30), the central rate of marriage for bachelors is

$$(bm)_x^h \doteq \frac{(bd)_x^h}{(bl)_x - \frac{1}{2}(bd)_x^h - \frac{1}{2}(bd)_x^d}. \qquad (19{\cdot}38)$$

Similarly, treating $(bd)_x^h$ as a negative decrement, we find that the force of mortality and the central rate of mortality for married men are

$$(h\mu)_x^d \doteq \frac{(hd)_{x-1}^d + (hd)_x^d}{2(hl)_x} \qquad (19{\cdot}39)$$

and

$$(hm)_x^d \doteq \frac{(hd)_x^d}{(hl)_x + \frac{1}{2}(bd)_x^h - \frac{1}{2}(hd)_x^d}. \qquad (19{\cdot}40)$$

It must be noted that the probability that a married man aged x will die within a year is not $(hd)_x^d/(hl)_x$ because $(hd)_x^d$ includes deaths among men who married between ages x and $x+1$ and therefore are not included in $(hl)_x$. For this probability reference must be made to a table representing the mortality of married men.

Also, an expression is sometimes required for the probability that a bachelor aged x will be alive and married at age $x+n$. Unless x is younger than the youngest age at which there are values of $(hl)_x$, this value will not be $(hl)_{x+n}/(bl)_x$ but

$$\frac{(hl)_{x+n} - (hl)_x \cdot {}_np_x}{(bl)_x},$$

where ${}_np_x$ is taken from a table representing the mortality of married men. This is because the $(hl)_{x+n}$ married men alive at age $x+n$ include the survivors of the $(hl)_x$ men who were alive and married at age x.

A combined mortality and disability table is similar to a combined mortality and marriage table. The numbers attaining age x are divided into active and disabled and the decrements are 'active persons dying', 'active persons becoming disabled' and 'disabled persons dying'. Disability is assumed to be permanent. Expressions for the force of disablement, the probability of becoming disabled within a year, and so on, may be obtained on the lines of formulae (19·36), (19·37), etc.

19·10. Selection

It is often found that the observed rates of decrement of members of a class (at least during the first few years of membership of the class) depend on the duration of their membership as well as on their age.

For example, the observed rates of withdrawal of a large group of employees of the same attained age who are members of a pension scheme are usually higher among those whose period of membership is short than among those who have been members of the scheme for some years. Notwithstanding this fact, multiple-decrement tables are not normally constructed on a select basis because the additional labour involved in constructing and using them would usually be disproportionate to any resulting gain in accuracy. A select multiple-decrement table, like a select mortality table, is merely a combination of a number of ordinary tables; the theory of such a table does not present any fresh difficulties and therefore it need not be discussed here.

Similarly, the observed rates of mortality of a group of disabled lives are usually relatively high when the duration of disability is short. The portion of a combined mortality and disability table which relates to disabled lives does not lend itself to construction on a select basis, because the number of entrants to the 'disabled' class at each age is determined by the number of decrements by disablement from the 'active' class.

EXAMPLES

Example 19·1

In a certain double-decrement table the probability of withdrawal, $(aq)_x^w$, is always equal to ·05. The probability $(aq)_{40}^d$ is equal to ·005 and the value of $(ad)_x^d$ is constant for all values of $x \geqslant 40$. Calculate the limiting age of the table.

Solution

We could proceed to construct the table age by age as follows:

x	$(al)_x$	$(ad)_x^w$	$(ad)_x^d$
40	100,000	5,000	500
41	94,500	4,725	500
42	89,275	...	...
...	...	...	...

However, this method would obviously be tedious and a more direct solution must be sought.

Now, using a radix of 100, we have

$$(al)_{41} = \cdot 95 \times 100 - \tfrac{1}{2},$$
$$(al)_{42} = \cdot 95^2 \times 100 - \tfrac{1}{2}(1 + \cdot 95),$$
$$\cdots\cdots\cdots$$
$$(al)_{40+t} = \cdot 95^t \times 100 - \frac{1}{2} \cdot \frac{1 - \cdot 95^t}{1 - \cdot 95}.$$

For $(al)_{40+t}$ to be equal to zero,

$$\cdot 95^t \times 100 = \frac{1 - \cdot 95^t}{2 \times \cdot 05}.$$
$$\therefore \quad \cdot 95^t \times 11 = 1.$$
$$\therefore \quad t \log \cdot 95 = \log \tfrac{1}{11}.$$
$$\therefore \quad t = \frac{-1 \cdot 04139}{\bar{1} \cdot 97772} = \frac{1 \cdot 04139}{\cdot 02228}$$
$$= 46 \cdot 7.$$

Hence the limiting age of the table is

$$40 + 47 = 87.$$

Example 19·2

A certain firm employs a large number of female workers between the ages of 25 and 30, and as part of their emoluments pays to an insurance company an annual premium (yearly in advance) to secure on behalf of each the following benefits:

(*a*) £25 on death in the service;

(*b*) £50 as a marriage dowry (if married before age 30);

(*c*) £75 on leaving the service at age 30.

Withdrawal is compulsory on marriage; no benefit is paid on withdrawal for any other cause.

Complete the following multiple-decrement table, being given the information that at each age the central rate of mortality is 1 per cent and the central rate of marriage 10 per cent, and calculate, using interest at 4 per cent, the net annual premium in respect of an entrant at age 25 to secure the benefits.

You are also given that at 4 per cent $v = \cdot 96154$, $v^2 = \cdot 92456$, $v^3 = \cdot 88900$, $v^4 = \cdot 85480$ and $v^5 = \cdot 82193$.

Age x	Number expected to attain age x in the service $(al)_x$
25	1,000
26	850
27	725
28	605
29	520
30	460

Solution

Let us denote marriage by the letter h.

The table is easily completed, because the number of deaths expected at each age is $\frac{1}{2}\{(al)_x + (al)_{x+1}\}(am)_x^d$, the number of marriages is

$$\tfrac{1}{2}\{(al)_x + (al)_{x+1}\}(am)_x^h$$

and the number of withdrawals is obtained by deducting the deaths and the marriages from $(al)_x - (al)_{x+1}$.

x	$(al)_x$	$\frac{1}{2}\{(al)_x + (al)_{x+1}\}$	Deaths $(ad)_x^d$	Marriages $(ad)_x^h$	Withdrawals $(ad)_x^w$
25	1,000	925	9	93	48
26	850	787·5	8	79	38
27	725	665	7	66	47
28	605	562·5	6	56	23
29	520	490	5	49	6
30	460				

To calculate the required annual premium, we proceed as follows:

x	$v^{x-25}(al)_x$	$v^{x-24}(ad)_x^d$	$v^{x-24}(ad)_x^h$
25	1,000·0	8·7	89·4
26	817·3	7·4	73·0
27	670·3	6·2	58·7
28	537·8	5·1	47·9
29	444·5	4·1	40·3
(Total)	(3,469·9)	(31·5)	(309·3)
30	378·1		

Then, if P is the required premium and if the death and marriage benefits are payable immediately on death or marriage,

$$P \times 3{\cdot}4699 = 25 \times {\cdot}0315 \times 1{\cdot}02 + 50 \times {\cdot}3093 \times 1{\cdot}02 + 75 \times {\cdot}3781.$$

$$\therefore \quad P = \frac{{\cdot}803 + 15{\cdot}774 + 28{\cdot}358}{3{\cdot}4699}$$

$$= \frac{44{\cdot}935}{3{\cdot}4699} = 12{\cdot}950 = £12\ 19s.$$

Rough check

Maximum amount payable in premiums = £65 approx. This appears reasonable in view of the benefits.

Example 19·3

If in a combined mortality and marriage table $(bl)_x$ = the number of bachelors expected to attain precise age x and $(hl)_x$ = the number of married men expected to attain precise age x, then, given the rate of interest and the values for all ages of

$$(bD)_x = v^x(bl)_x; \quad (bN)_x = \sum_{t=0}^{\infty} (bD)_{x+t};$$

$$(hD)_x = v^x(hl)_x; \quad (hN)_x = \sum_{t=0}^{\infty} (hD)_{x+t};$$

a_x = the value of an immediate annuity payable during the whole future lifetime of a married man aged x:

show how you would calculate accurately the annual premium, limited to 35 payments, chargeable during bachelorhood in the case of a bachelor aged 25 for each of the following:

(*a*) an assurance payable at his death whether he shall have married or not:

(*b*) a deferred annuity payable from age 60 during his lifetime but only if he shall have married before that age.

Solution

(*a*) We require first an annuity payable during the remainder of the lifetime of a bachelor aged 25.

The value of an annuity-due for life to all the $(bl)_{25}+(hl)_{25}$ persons aged exactly 25 is

$$\{(bl)_{25}+(hl)_{25}\}\frac{(bN)_{25}+(hN)_{25}}{(bD)_{25}+(hD)_{25}}=\frac{(bN)_{25}+(hN)_{25}}{v^{25}}.$$

The value of an annuity-due for life to all the $(hl)_{25}$ who are married is

$$(hl)_{25}\,\ddot{a}_{25}.$$

Therefore, the value of an annuity-due for life to all the $(bl)_{25}$ who are bachelors is

$$\frac{(bN)_{25}+(hN)_{25}}{v^{25}}-(hl)_{25}\,\ddot{a}_{25}.$$

To obtain the value of an annuity-due to one bachelor we divide this expression by $(bl)_{25}$, which gives

$$\frac{(bN)_{25}+(hN)_{25}-(hD)_{25}\,\ddot{a}_{25}}{(bD)_{25}}.$$

To obtain the single premium for the assurance we deduct 1 from this annuity and enter conversion tables. To obtain the annual premium we multiply the single premium by

$$\frac{(bD)_{25}}{(bN)_{25}-(bN)_{60}}.$$

(*b*) By means of the relationship $a_x=vp_x\ddot{a}_{x+1}$, construct a table of p_x or $\log p_x$ from the given values of a_x. Hence, obtain (most conveniently by logarithms) ${}_{35}p_{25}$, the chance that a married man aged 25 will survive to age 60.

The number of married men living at age 60 who were bachelors at age 25 is

$$(hl)_{60}-{}_{35}p_{25}\,(hl)_{25}$$

and the desired annual premium is, therefore,

$$\frac{(hD)_{60}-v^{35}\,{}_{35}p_{25}\,(hD)_{25}}{(bN)_{25}-(bN)_{60}}\,a_{60}.$$

The danger of granting such contracts on individual lives is obvious.

CHAPTER 20

THE MULTIPLE-DECREMENT TABLE AND ITS RELATED SINGLE-DECREMENT TABLES

20·1. Introduction

In respect of every multiple-decrement table with, say, r modes of decrement we may imagine a set of r single-decrement tables which are related to it.

In one of these tables the mode of decrement is death; this table is therefore a mortality table. The other tables are similar in every way to mortality tables except that the mode of decrement is some other event such as withdrawal. It may seem at first that the idea of a group of lives who are being continually reduced in number by withdrawal, but not by death, is absurd and meaningless. The concept, however, is of considerable value, not only in the study of the theory of the multiple-decrement table but also, as will be seen later, in practical work.

It is therefore proposed in this chapter to study the model, consisting of the multiple-decrement table and its related single-decrement tables, to consider the interpretations to be placed upon it and to develop certain relationships between the functions of the multiple-decrement table and the functions of the related single-decrement tables.

20·2. Notation

The notation used in Great Britain for the related single-decrement tables is the same as that set out in Volume I of this book except that when it is desired to specify the mode of decrement this is done by adding a letter at the upper right-hand corner of the main symbol. For example, the symbols l_x^α, μ_x^α, etc., are used in theoretical work to represent functions in a table in which the only mode of decrement is α. For convenience, the notation used for multiple-decrement and single-decrement tables is set out in tabular form below.

TABLE 20·1

Function	Symbol in the multiple-decrement table		Symbol in the single-decrement table
	All decrements combined	Decrement α	Decrement α
Number expected to survive* to age x	$(al)_x$	—	l_x^α
Force of decrement	$(a\mu)_x$	$(a\mu)_x^\alpha$	μ_x^α
Expected number of decrements between ages x and $x+1$	$(ad)_x$	$(ad)_x^\alpha$	d_x^α
Probability or rate of decrement	$(aq)_x$	$(aq)_x^\alpha$	q_x^α
Probability of survival* from age x to age $x+t$	${}_t(ap)_x$	—	${}_tp_x^\alpha$

When the mode of decrement is β, γ, etc., the letter α will be replaced by β, γ, etc.

20·3. 'Rate', 'probability' and 'force'

It is important to understand clearly the meanings which are attached to the words 'rate', 'probability' and 'force' in connexion with the multiple-decrement or single-decrement table.

Just as q_x in a mortality table is referred to as the rate of mortality, so q_x^α in a single-decrement table in which the mode of decrement is α is referred to as the rate of decrement. Thus q_x^w is the rate of withdrawal, q_x^i is the rate of ill-health retirement, and so on. In Volume I of this book, q_x was defined alternatively as being 'the rate of mortality at age x' and 'the probability that (x) will die within a year'. Although the word 'probability' can be used in this way when we are dealing with a class of lives which is assumed to be subject to no other mode of decrement than death, the general use of the word as an alternative to 'rate' is undesirable. For example, in connexion with a single-decrement table in which the decrement is withdrawal, we cannot say that q_x^w is the probability that a life aged x, taken at random from an indefinitely large number

* In connexion with the multiple-decrement table, the word 'survive' is used here as an abbreviation for 'remain a member of the class to which the $(al)_x$ column of the table relates'; the word 'survival' is used similarly.

of lives whose experience is assumed to be represented by the table, will withdraw before attaining age $x+1$, because although the mortality element has been excluded from the model it cannot be eliminated from actual experience. (Should it be necessary to translate into words a symbol such as ${}_nq_x^\alpha$, where n is not equal to 1, it may sometimes be difficult to avoid the use of the word 'probability'. If the mode of decrement is not death, it should be made clear that the probability relates only to a unit in the model and not to a life.)

In a multiple-decrement table, each of the ratios $(aq)_x^\alpha$, $(aq)_x^\beta$, etc., is referred to as a probability. The use of the word is obviously legitimate, provided that it is applied to a life who is taken at random from an indefinitely large number of lives who are assumed to experience, as a whole, the mortality and other decrements on which the table is based. These functions may be referred to also as rates. In order to avoid confusion by the use of the term 'rate' for both q_x^α and $(aq)_x^\alpha$ the former is sometimes referred to as the 'independent rate' and the latter as the 'dependent rate'. The reason for this terminology will be explained in a later section of this chapter.

We have already seen that the expressions for the forces of decrement involve differential coefficients; they therefore exist only in the model. Ratios obtained from an experience necessarily relate to finite intervals; if, however, they take the form ${}_h(aq)_x^\alpha$, where h is small, they may be regarded as providing an approximation to the forces of decrement, since

$$(a\mu)_x^\alpha = \lim_{h\to 0} \frac{{}_h(aq)_x^\alpha}{h}.$$

Strictly speaking, the central rates also exist only in the model, but, as has already been pointed out, when a mortality or other investigation is being conducted the ratios which are obtained from the data often resemble central rates.

20·4. Selective and non-selective decrements

For a proper study of the relationships between the multiple-decrement table and its related single-decrement tables, it is essential to distinguish between 'selective' and 'non-selective' decrements. A selective decrement may be defined as one which takes from the class some members who are not random so far as another decrement is concerned; a decrement is non-selective if the members whom it takes from the class are random in respect of all the other decrements.

As an illustration of a selective decrement, let us suppose that a large organization institutes a pension scheme for its employees and let us assume that, after a certain age, there are no withdrawals from service before the attainment of the normal pension age except by ill-health retirement. It will almost certainly be found that the observed mortality of the ill-health pensioners is heavier, especially during the first few years after retirement, than the mortality in a single-decrement table constructed to reflect the combined mortality experience of active employees and pensioners. In other words, the members who leave the class by ill-health retirement are not random so far as mortality is concerned. In these circumstances ill-health retirement is stated to be a selective decrement.

In practice, almost every decrement is selective to a greater or less degree and an example of a completely non-selective one is therefore not easy to find. However, a good illustration is provided by the Victory Bond policies that were issued by British life offices after the war of 1914–18. Victory Bonds were issued at a discount by the Government and were redeemable by drawings at par each year; all the bonds in force immediately before any drawing took place had an equal chance of being drawn. Each Victory Bond policy was issued in respect of an individual bond, the number of the bond being specified in the policy. Under one form of policy the bond was transferred to the assured or his representatives on the survival of the life assured for a selected term or on his previous death. If the bond was drawn for repayment during the selected term and during the lifetime of the life assured, the redemption sum of the bond was transferred immediately and the policy terminated. In order to calculate the appropriate premium for such a policy a double-decrement table could have been constructed with two modes of decrement, death and 'drawing of a bond'. Since the bonds were drawn at random, the decrement 'drawing of a bond' would have been non-selective. Similarly, the lives assured who died from time to time would have been random as regards their prospects of having their bonds drawn, and mortality also would have been a non-selective decrement.

Notwithstanding the rarity of completely non-selective decrements, it is often convenient to assume that all the decrements in a table are non-selective. The error introduced by this assumption may be small in relation to the effect of other influences such as approximations, inadequacy of the data, uncertainty as to the suitability of the table to the problem in hand, and so on. The estimation of the relative magnitude of these various influences is a matter that calls for skill and judgment and it will not be discussed in this book.

20·5. Selective decrements—an alternative view

It will be clear from the Victory Bond illustration in the preceding section that when one decrement (say decrement α) is independent of any attribute of the lives, then decrement α is non-selective; if there are only two modes of decrement in the table, the other decrement also will be non-selective. The Victory Bond example, however, is exceptional. Usually each decrement in the table is influenced by one or more attributes of the lives and it is very unlikely that the attributes which influence decrement α will be completely independent of those which influence the others. However, if in any table this complete independence did exist, then decrement α would be non-selective and if there were only two decrements in the table the other decrement also would be non-selective.

It follows that we could define the terms 'selective' and 'non-selective' by saying that if the circumstances or attributes influencing decrement α are independent of those influencing the other decrement or decrements, then decrement α is non-selective, but that if this independence does not exist decrement α is selective. This definition is in terms of the causes of the selection whereas the definition in section 4 of this chapter is in terms of the effects which the selection produces. We can often satisfy ourselves by general reasoning, as in the illustration of the pension scheme above, that a decrement will be selective, or occasionally, as in the case of the Victory Bond policies, that it will be non-selective, but in order to determine the extent of the selective influence we must rely on a study of actual experience of a group of lives. For this reason it is better to define the terms 'selective' and 'non-selective' as in section 20·4.

20·6. The interpretation of the single-decrement tables when the decrements are non-selective

In order that the model, consisting of the multiple-decrement table and its related single-decrement tables, may be of practical value, it is important to be able to interpret it in terms of experience.

In considering this problem let us confine our attention to a double-decrement table in which the modes of decrement are mortality and withdrawal and let us assume that these decrements are non-selective. The extension of the argument to a model in which there are three or more modes of decrement will be sufficiently obvious. The consideration of a model which is intended to represent the operation of selective decrements will be deferred to a later section of the chapter.

We have already seen in Chapter 19 that $(al)_x$ in such a table can be regarded as corresponding to an empirical sequence $(an)_x$, the number of persons surviving to age x as members of a designated class of persons who are subject to the two modes of decrement, death and withdrawal. We have now to consider how we are to define the empirical sequences (say n_x^d and n_x^w) to which l_x^d and l_x^w, the fundamental functions of the related single-decrement tables, are to be regarded as corresponding. We could, of course, say that each sequence represents the number of persons that would be observed to survive to age x (out of a large number, n_α^d or n_α^w, coming under observation at age α) if one mode of decrement were operating alone, but this definition lacks precision.

So far as n_x^d (corresponding to l_x^d) is concerned, the problem presents no difficulty. The $(an)_x$ persons corresponding to the $(al)_x$ of the double-decrement table possess two sets of attributes, of which one set influences mortality and the other set influences withdrawal; these two sets of attributes, which may be referred to as mortality attributes and withdrawal attributes respectively, are assumed to be completely independent of each other. We have only to assume, therefore, that for every value of x the n_x^d group, considered as a whole, is indistinguishable from the $(an)_x$ group, also considered as a whole, in respect of the mortality attributes, but that the n_x^d persons differ from the $(an)_x$ persons in possessing withdrawal attributes which are so constituted that withdrawal cannot occur.

It should be appreciated that the individual persons in the n_x^d and $(an)_x$ groups (like the n_x individuals considered in Chapter 1 of Volume I) are not necessarily to be regarded as identical in respect of all the attributes which influence mortality. All that is stipulated is that the n_x^d group, considered as a whole, is indistinguishable in respect of those attributes from the $(an)_x$ group, also considered as a whole.

When we come to consider the n_x^w persons corresponding to the l_x^w of the table in which the only decrement is withdrawal, we may employ a similar argument. We may assume that for every value of x the n_x^w group, considered as a whole, is indistinguishable from the $(an)_x$ group, also considered as a whole, in respect of the withdrawal attributes and that the n_x^w persons differ from the $(an)_x$ persons in possessing mortality attributes which are so constituted that death cannot occur. This assumption is of course incapable of realization in practice.

The device described above not only helps us to visualize a group of persons who are subject to a single decrement other than mortality, but provides a relationship, in terms of experience, between the multiple-decrement table and each related single-decrement table. This relationship consists in the fact that, so far as the attributes influencing decrement α are concerned, the n_x^α group corresponding to the l_x^α of the single-decrement table is indistinguishable from the $(an)_x$ group corresponding to the $(al)_x$ of the multiple-decrement table.

If we construct two models (each consisting of a double-decrement table with its two related single-decrement tables) to represent the mortality and withdrawal experience of two groups of lives which are indistinguishable from each other in respect of mortality attributes, but differ from each other in respect of withdrawal attributes, the rates of mortality q_x^d in the two models will be identical.

The probabilities $(aq)_x^d$, however, will differ, since the numbers exposed to death will be decreasing at different rates in the two tables. In other words, the rates of mortality q_x^d will be independent of the withdrawal rates but the probabilities $(aq)_x^d$ will not. Similarly, if the two groups are indistinguishable in respect of withdrawal attributes, but differ in respect of mortality attributes, the rates of withdrawal q_x^w will be independent of the mortality rates but the probabilities $(aq)_x^w$ will not. It is for this reason that the rates in the single-decrement tables and the probabilities in the multiple-decrement table are sometimes referred to as independent rates and dependent rates respectively.

20·7. The identity of the forces

The force $(a\mu)_x^\alpha$ may be regarded as an instantaneous measure of the rate at which $(al)_x$ is decreasing as a result of the operation of the decrement α when other decrements β, γ, ... also are operating. If the decrements are non-selective the force μ_x^α may be regarded as an instantaneous measure of the rate at which l_x^α is decreasing when the decrement α is operating alone in the manner visualized in the previous section. Since it is assumed that the attributes influencing decrements β, γ, ... are independent of those influencing decrement α, it follows that the two forces $(a\mu)_x^\alpha$ and μ_x^α are identical,*

i.e. that

$$\left.\begin{aligned}(a\mu)_x^\alpha &\equiv \mu_x^\alpha.\\ \text{Similarly}\qquad (a\mu)_x^\beta &\equiv \mu_x^\beta,\end{aligned}\right\} \tag{20·1}$$

and so on.

The identity of the forces of decrement is the fundamental connecting link between the multiple-decrement table and its related single-decrement tables. It is of great importance and forms the basis of the theory of the multiple-decrement table and its related single-decrement tables.

20·8. The practical usefulness of the model

Provided that all the decrements are non-selective or that we are satisfied that the error involved in assuming them to be non-selective is negligible, the model, consisting of the multiple-

* Those who find this argument unconvincing are referred to section 20·10 for a demonstration which is based on the assumption that the decrements are selective but is equally valid when they are non-selective.

decrement table and its related single-decrement tables, is a very powerful actuarial tool because:

(i) we can convert a multiple-decrement table with, say, r modes of decrement into another table with $r+1$ decrements by combining another single-decrement table with the r single-decrement tables which are related to the original table,

(ii) we can convert the original table into one with $r-1$ decrements by omitting one of the r single-decrement tables,

(iii) we can convert the original table into another one with r decrements by replacing one of the related single-decrement tables by another single-decrement table in which the mode of decrement is the same but the rates of decrement are different,

(iv) given several different models, each consisting of a multiple-decrement table with its related single-decrement tables, we can construct a new multiple-decrement table by taking one single-decrement table from the first model, another single-decrement table from the second model, and so on, and combining them,

(v) we can regard the l_x of an ordinary mortality table as being l_x^d for the purpose of combining it with other single-decrement tables in order to form a new multiple-decrement table.

*20·9. The interpretation of the single-decrement tables when the decrements are selective

The interpretation of the single-decrement tables which was set out in section 6 of this chapter is open to the objection that it is based on the assumption that the decrements are non-selective—an assumption which is rarely fulfilled in practice. It is therefore proposed in the present section to give an interpretation which applies when the decrements are selective (and also when they are non-selective). In the following section we shall see that the identities

$$(a\mu)_x^\alpha \equiv \mu_x^\alpha,$$
$$(a\mu)_x^\beta \equiv \mu_x^\beta,$$

and so on, are applicable to this wider interpretation.

Before we attempt to interpret the single-decrement tables, regarded as related to the multiple-decrement table, it will be helpful to consider each single-decrement table by itself. Let us

* See Preface for significance of this asterisk.

take first the withdrawal table, in which the fundamental function is l_x^w. What meaning are we to assign to the corresponding sequence n_x^w when the decrements are selective?

We may imagine that there is an indefinitely large number, $(aN)_\alpha$, of persons all attaining exact age α during a short period of time, that they are subject to two modes of decrement, death and withdrawal, and that each of them is to be kept under observation until he dies or withdraws. From this indefinitely large number a smaller (but still large) number, n_α^w, of persons is taken at random and each of them is kept under observation. Whenever one of these n_α^w persons withdraws (say at age x), his withdrawal has the effect of depleting n_x^w by one, but whenever one of them dies he is immediately replaced by another person aged x, taken at random from the $(aN)_x$ survivors of the original $(aN)_\alpha$ persons. It will be seen that by this replacement device it is possible to produce the effect of a group of n_α^w persons who are subject to only one mode of decrement, withdrawal.

The other single-decrement table, in which the mode of decrement is death, is of course a mortality table and its fundamental function l_x^d could be regarded (like the l_x of the mortality table considered in Volume I, Chapter 1) as corresponding to the survivors of a group of persons who are kept under observation until death. For our present purpose, however, it is more satisfactory to adopt the device of the preceding paragraph and to think of n_α^d persons taken at random from $(aN)_\alpha$ persons who are subject to two modes of decrement, death and withdrawal, and to imagine that whenever one of the n_α^d persons withdraws (say at age x) he is immediately replaced by another person aged x, taken at random from the $(aN)_x$ survivors of the original large number $(aN)_\alpha$.

Let us now consider the model as a whole. Let us suppose that from the indefinitely large number, $(aN)_\alpha$, of persons, three groups are taken at random, numbering $(an)_\alpha$, n_α^d and n_α^w respectively. Each of the $(an)_\alpha$ persons in the first group is kept under observation until he leaves either by death or by withdrawal. The number, $(an)_x$, of persons remaining in this group corresponds to the function $(al)_x$ in the double-decrement table; the numbers of persons leaving this first group by death and withdrawal in any year of age x to $x+1$ correspond to the functions $(ad)_x^d$ and $(ad)_x^w$. By the replacement device described above, the number in the second group of n_α^d persons is depleted by deaths only; the number, n_x^d, of persons remaining in this group corresponds to the function l_x^d in the single-decrement mortality table. Similarly, by the replacement method, the number in the third group of n_α^w persons is depleted by withdrawals only and the number, n_x^w, of persons remaining in this group corresponds to the function l_x^w.

It will be seen that this device, like that of section 20·6, not only provides an analogy, in terms of experience, for the single-decrement tables but also indicates, in terms of experience, the nature of the relationship between the double-decrement table and its related single-decrement tables. This relationship is provided by the fact that every original member of the groups of n_α^d and n_α^w persons, and every member subsequently introduced into those groups in order to replace withdrawals and deaths respectively, is taken at random from the same $(aN)_\alpha$ persons from whom the $(an)_\alpha$ persons are taken. The argument can easily be extended to a table with more than two decrements.

*20·10. The identity of the forces when the decrements are selective

It is now proposed to show that, provided that the model is interpreted as described in the preceding section, the identity of the forces applies even when the decrements are selective.

For this purpose it is convenient to regard the period from age α to age ω (ω being the youngest age at which $(aN)_x$ is zero) as being divided into a large number, n, of equal intervals each equal to h, so that $\omega-\alpha=nh$, and to regard the suffix x as applying only to the n ages α, $\alpha+h$, $\alpha+2h$, ..., $\omega-h$. Let ${}_h(aq)_x^d$ and ${}_h(aq)_x^w$ be the observed probabilities of death and withdrawal between ages x and $x+h$ among the $(an)_x$ persons and let ${}_hq_x^d$ and ${}_hq_x^w$ be the observed ratios by which the n_x^d and n_x^w persons respectively are depleted in number between ages x and $x+h$. (It is assumed that $(an)_x$, n_x^d and n_x^w are sufficiently large for random variations to be negligible.)

Let us now make, for all the n values of x, the following two alternative assumptions as to the manner in which the deaths and withdrawals that occur during the interval between ages x and $x+h$ among the $(aN)_x$ persons are distributed over this interval:

(i) that all the deaths occur between ages x and $x+k$ $(k<h)$ and all the withdrawals occur between ages $x+k$ and $x+h$,

(ii) that all the withdrawals occur between ages x and $x+k$ and all the deaths occur between ages $x+k$ and $x+h$.

On assumption (i), the ratio by which the $(an)_x$ persons are depleted in number as a result of the deaths between ages x and $x+k$ is the same as the ratio by which the n_x^d persons are depleted (because both groups are random samples from the same large number $(aN)_x$), i.e.

$$_h(aq)_x^d = {}_hq_x^d. \tag{20·2}$$

The $(an)_x$ persons are therefore reduced in number, between ages x and $x+k$, to

$$(an)_x(1-{}_hq_x^d).$$

The ratio by which these remaining persons are depleted, as a result of the withdrawals which occur between ages $x+k$ and $x+h$, is the same as the ratio by which the n_x^w persons are depleted, i.e. it is equal to ${}_hq_x^w$. It follows that the number of withdrawals occurring, among the $(an)_x$ persons, between ages $x+k$ and $x+h$ is

$$(an)_x(1-{}_hq_x^d)\,{}_hq_x^w.$$

But this number of withdrawals is also

$$(an)_x \cdot {}_h(aq)_x^w.$$

$$\therefore \quad {}_h(aq)_x^w = (1-{}_hq_x^d)\,{}_hq_x^w$$

or

$$_h(aq)_x^w = {}_hq_x^w - {}_hq_x^d \cdot {}_hq_x^w. \tag{20·3}$$

By a similar process of reasoning, it is easy to show that on assumption (ii)

$$_h(aq)_x^w = {}_hq_x^w \tag{20·4}$$

and

$$_h(aq)_x^d = {}_hq_x^d - {}_hq_x^d \cdot {}_hq_x^w. \tag{20·5}$$

Let us now assume that the number, n, of intervals into which the period between ages α and ω is divided is gradually increased, so that h gradually decreases, nh being constant. As h decreases, the term ${}_hq_x^d \cdot {}_hq_x^w$ which occurs on the right-hand side of formulae (20·3) and (20·5) becomes less and less significant by comparison with the other terms. We cannot regard h as becoming indefinitely small when we are dealing with groups of lives because, even if the groups are large, the deaths and withdrawals among them must occur at finite intervals. This limitation, however, does not apply to the model. Let us therefore imagine that we have two models, one designed to represent assumption (i) and the other designed to represent assumption (ii). Equations (20·2) and (20·3) will apply to the first model and equations (20·4) and (20·5) will apply to the second. If now n is made larger and larger until it tends to infinity and h tends to zero, the two models will approximate more and more closely to each other until, in the limit, they become identical with each other and with the normal model. When this limiting stage is reached, we shall have from equations (20·2) and (20·5)

$$\lim_{h\to 0} \frac{{}_h(aq)_x^d}{h} = \lim_{h\to 0} \frac{{}_hq_x^d}{h}$$

or

$$(a\mu)_x^d = \mu_x^d, \tag{20·6}$$

and from equations (20·3) and (20·4)

$$\lim_{h\to 0} \frac{{}_h(aq)_x^w}{h} = \lim_{h\to 0} \frac{{}_hq_x^w}{h}$$

or

$$(a\mu)_x^w = \mu_x^w. \tag{20·7}$$

This argument, like that of section 20·9, can be extended to a table with more than two decrements.

We see therefore that if the model is to correspond to experience in the manner described in section 20·9, the identity of the forces of decrement is a necessary connecting link between the multiple-decrement table and the related single-decrement tables.

20·11. Construction of the multiple-decrement table from the related single-decrement tables

It will be apparent from what has been said in section 20·8 that it is frequently necessary to construct a multiple-decrement table from the related single-decrement tables, and vice versa. It is proposed in this section to consider the problem of constructing a multiple-decrement table from the related single-decrement tables.

In theory (subject to the point mentioned on page 5) the multiple-decrement table is fully defined if the radix is given and the forces

$(a\mu)_x^\alpha \equiv \mu_x^\alpha$, etc., are known for all values of x. In practice, however, the only convenient way of constructing the table is to multiply the radix by the values of $(aq)^\alpha$, $(aq)^\beta$, ... for the youngest age in the table, so obtaining $(ad)^\alpha$, $(ad)^\beta$, ... for that age, to deduct these decrements from the radix, so obtaining the value of (al) for the next age in the table, and so on. The information that is available in respect of any one of the related single-decrement tables (say the table in respect of decrement α) will usually consist of the numerical values of l_x^α, d_x^α and q_x^α for integral values of x; μ_x^α also, if given at all, will usually be known for integral values of x only. It follows that the identity $(a\mu)_x^\alpha \equiv \mu_x^\alpha$, although of great theoretical importance, is of little value for constructing the multiple-decrement table from the related single-decrement tables. What is required for this purpose is an approximate formula giving $(aq)_x^\alpha$ in terms of q_x^α, q_x^β, etc.

Before we approach this problem, it will be convenient to consider the simpler one of constructing the $(al)_x$ and $(ad)_x$ columns only, without dividing $(ad)_x$ into its component parts $(ad)_x^\alpha$, $(ad)_x^\beta$, etc. Since

$$(a\mu)_x = (a\mu)_x^\alpha + (a\mu)_x^\beta + \dots$$
$$\equiv \mu_x^\alpha + \mu_x^\beta + \dots,$$

it follows that

$$_n(ap)_x = \exp\left\{-\int_0^n (a\mu)_{x+t}\,dt\right\}$$
$$= \exp\left\{-\int_0^n \mu_{x+t}^\alpha\,dt\right\} \times \exp\left\{-\int_0^n \mu_{x+t}^\beta\,dt\right\} \times \dots,$$

i.e.

$$_n(ap)_x = {}_np_x^\alpha \times {}_np_x^\beta \times \dots. \tag{20·8}$$

This formula is analogous to the product rule of probability theory, which applies when the probabilities $_np_x^\alpha$, etc., are independent.

When $n=1$, formula (20·8) can be written

$$1-(aq)_x = (1-q_x^\alpha)(1-q_x^\beta)\dots$$

or

$$(aq)_x = \Sigma q_x^\alpha - \Sigma[q_x^\alpha q_x^\beta] + \Sigma[q_x^\alpha q_x^\beta q_x^\gamma] - \dots. \tag{20·9}$$

By means of this formula, if values of q_x^α, etc., are given, $(aq)_x$ can be calculated and hence the $(al)_x$ and $(ad)_x$ columns can be constructed.

When there are only two modes of decrement, formula (20·9) reduces to

$$(aq)_x = q_x^\alpha + q_x^\beta - q_x^\alpha q_x^\beta. \tag{20·10}$$

It will be seen that this formula is consistent with formulae (20·2) and (20·3), and also with formulae (20·4) and (20·5), when $h=1$. This is to be expected, since the formula has been deduced without any assumption as to the manner in which the decrements are distributed over the year of age x to $x+1$.

Coming now to the more usual problem of constructing the complete multiple-decrement table from the related single-decrement tables, we have from formula (19·26)

$$(aq)_x^\alpha = \int_0^1 {}_t(ap)_x \,(a\mu)_{x+t}^\alpha \,dt$$

$$= \int_0^1 \frac{{}_t(ap)_x}{{}_tp_x^\alpha}\, {}_tp_x^\alpha \mu_{x+t}^\alpha \,dt.$$

If we assume that ${}_tp_x^\alpha \mu_{x+t}^\alpha$ is constant when $0 \leqslant t < 1$, i.e. that the decrements in the single-decrement table α are distributed uniformly over the year of age, this relationship becomes

$$(aq)_x^\alpha = q_x^\alpha \int_0^1 \frac{{}_t(ap)_x}{{}_tp_x^\alpha}\,dt$$

$$= q_x^\alpha \int_0^1 {}_tp_x^\beta \cdot {}_tp_x^\gamma \cdot {}_tp_x^\delta \cdot \ldots \,dt.$$

If we assume that the decrements in *each* single-decrement table are distributed uniformly over the year of age, it follows that when $k<1$

$${}_kq_x^\beta = kq_x^\beta, \quad \text{etc.}$$

Hence
$$(aq)_x^\alpha = q_x^\alpha \int_0^1 (1 - tq_x^\beta)(1 - tq_x^\gamma)\ldots\,dt$$

$$= q_x^\alpha \int_0^1 \{1 - t\Sigma q_x^\beta + t^2\Sigma[q_x^\beta q_x^\gamma] - \ldots\}\,dt,$$

where the summations apply to all modes of decrement except α.

$$\therefore \quad (aq)_x^\alpha = q_x^\alpha\{1 - \tfrac{1}{2}\Sigma q_x^\beta + \tfrac{1}{3}\Sigma[q_x^\beta q_x^\gamma] - \ldots\}. \tag{20·11}$$

When there are only two modes of decrement, formula (20·11) becomes

$$\left.\begin{aligned} (aq)_x^\alpha &= q_x^\alpha(1 - \tfrac{1}{2}q_x^\beta), \\ (aq)_x^\beta &= q_x^\beta(1 - \tfrac{1}{2}q_x^\alpha). \end{aligned}\right\} \tag{20·12}$$

whence

These formulae are exact on the assumption that the decrements in each single-decrement table are distributed uniformly over the year of age. Formulae on other assumptions as to the distribution of the decrements will be developed in section 13 of this chapter.

It sometimes happens that, when the related single-decrement tables are based on data obtained from investigations, reasonable approximations to the values of m_x^α, m_x^β, etc., are available. Now the difference between $(am)_x^\alpha$ and m_x^α is usually small. The relationships

$$(am)_x^\alpha = \frac{\int_0^1 (al)_{x+t}(a\mu)_{x+t}^\alpha\,dt}{\int_0^1 (al)_{x+t}\,dt}$$

and

$$m_x^\alpha = \frac{\int_0^1 l_{x+t}^\alpha \mu_{x+t}^\alpha\,dt}{\int_0^1 l_{x+t}^\alpha\,dt}$$

show (since $(a\mu)_x^\alpha \equiv \mu_x^\alpha$) that the central rates are weighted means of the forces, the weights being proportional to $(al)_{x+t}$ and l_{x+t}^α respectively. As the weights make very little difference to the means, it follows that

$$(am)_x^\alpha \doteq m_x^\alpha. \tag{20·13}$$

This result is apparent also from the relationships

$$(am)_x^\alpha \doteq (a\mu)_{x+\frac{1}{2}}^\alpha \equiv \mu_{x+\frac{1}{2}}^\alpha \doteq m_x^\alpha.$$

Writing m_x^α for $(am)_x^\alpha$ in formula (19·33), we have

$$(aq)_x^\alpha \doteq \frac{m_x^\alpha}{1+\frac{1}{2}(m_x^\alpha + m_x^\beta + \dots)} \tag{20·14}$$

and the multiple-decrement table can be constructed.

When only two modes of decrement are involved formula (20·14) reduces to

$$(aq)_x^\alpha \doteq \frac{m_x^\alpha}{1+\frac{1}{2}(m_x^\alpha + m_x^\beta)}. \tag{20·15}$$

However, if the decrements in each single-decrement table are distributed uniformly over the year of age and only two modes of decrement are involved,

$$q_x^\alpha = \frac{m_x^\alpha}{1+\frac{1}{2}m_x^\alpha} \quad \text{and} \quad (1-\tfrac{1}{2}q_x^\beta) = \frac{1}{1+\frac{1}{2}m_x^\beta}.$$

Substituting these values in formula (20·12) we have

$$(aq)_x^\alpha = \frac{m_x^\alpha}{(1+\frac{1}{2}m_x^\alpha)(1+\frac{1}{2}m_x^\beta)}. \tag{20·16}$$

This formula, being exact when the decrements in each single-decrement table are distributed uniformly over the year of age, is generally to be preferred to formula (20·15).

20·12. Construction of a related single-decrement table from the multiple-decrement table

In order to construct a related single-decrement table from the multiple-decrement table we normally require an expression for q^α in terms of $(aq)^\alpha$, $(aq)^\beta$, etc. If we first limit our consideration to two modes of decrement and assume that the decrements in the two single-decrement tables are distributed uniformly over the year of age, we have from formula (20·12)

$$q_x^\alpha = \frac{(aq)_x^\alpha}{1-\frac{1}{2}q_x^\beta}. \tag{20·17}$$

As a first approximation to q_x^α we can replace q_x^β by $(aq)_x^\beta$, whence

$$q_x^\alpha \doteq \frac{(aq)_x^\alpha}{1-\frac{1}{2}(aq)_x^\beta}. \tag{20·18}$$

This formula may be written

$$q_x^\alpha \doteq (aq)_x^\alpha\{1+\tfrac{1}{2}(aq)_x^\beta+\tfrac{1}{4}[(aq)_x^\beta]^2+\dots\}$$

or, if the decrements are not large,

$$q_x^\alpha \doteq (aq)_x^\alpha\{1+\tfrac{1}{2}(aq)_x^\beta\}. \tag{20·19}$$

To obtain a more accurate approximation (on the same assumption as to the distribution of the decrements) we replace q_x^α by $(aq)_x^\alpha$ in the formula

$$q_x^\beta = \frac{(aq)_x^\beta}{1-\frac{1}{2}q_x^\alpha}$$

and then substitute this value of q_x^β in formula (20·17). This gives

$$q_x^\alpha \doteq \frac{(aq)_x^\alpha}{1-\dfrac{\frac{1}{2}(aq)_x^\beta}{1-\frac{1}{2}(aq)_x^\alpha}},$$

whence

$$q_x^\alpha \doteq (aq)_x^\alpha \frac{1-\frac{1}{2}(aq)_x^\alpha}{1-\frac{1}{2}(aq)_x^\alpha-\frac{1}{2}(aq)_x^\beta}$$

or

$$q_x^\alpha \doteq (aq)_x^\alpha \frac{1-\frac{1}{2}(aq)_x^\alpha}{1-\frac{1}{2}(aq)_x}. \tag{20·20}$$

An accurate expression (still on the assumption that the decrements in each single-decrement table are distributed uniformly over the year of age) may be obtained as follows. Formula (20·12) may be written

$$(aq)_x^\alpha = q_x^\alpha \left\{1 - \frac{1}{2} \cdot \frac{(aq)_x^\beta}{1-\frac{1}{2}q_x^\alpha}\right\},$$

which gives a quadratic equation for q_x^α. Solving this equation and expanding by the binomial theorem we can show that

$$q_x^\alpha = \alpha\{1 + \tfrac{1}{2}\beta + \tfrac{1}{4}(\alpha\beta+\beta^2) + \tfrac{1}{8}(\alpha^2\beta + 3\alpha\beta^2 + \beta^3) + \ldots\}, \tag{20·21}$$

where α is written for $(aq)_x^\alpha$ and β for $(aq)_x^\beta$.

Formula (20·21) is not in a very convenient form for practical application. Its chief usefulness lies in the fact that it provides a measure of the degree of accuracy of other formulae. For example, formulae (20·18) and (20·20) can be converted, with the aid of the binomial theorem, into the form

$$q_x^\alpha = \alpha(1 + \tfrac{1}{2}\beta + \ldots).$$

If this is done it is easily seen that formula (20·19) is less accurate than formula (20·18) and that formula (20·20) is very nearly exact. In the same way it can be shown that the empirical formula

$$q_x^\alpha = (aq)_x^\alpha \frac{1-\frac{1}{2}(aq)_x^\beta}{1-(aq)_x^\beta}$$

is usually more accurate than formula (20·18).

Formula (20·19) has the advantage that it is the simplest to apply arithmetically and it gives sufficiently accurate results for most practical purposes. It should be remembered that the assumption of uniform distribution of decrements in the single-decrement tables introduces an error of unknown magnitude which may well be greater than the discrepancy between an approximate formula and formula (20·21). A practical point to be borne in mind is that $(aq)_x^\alpha$ and $(aq)_x^\beta$ are often so small in relation to unity that, unless it is desired to retain a large number of significant figures in the table, the differences between the numerical results of different formulae will be negligible.

Although formula (20·20) has been developed on the assumption that two modes of decrement only are involved it is equally applicable when there are any number of modes of decrement, because the decrement β can be assumed to represent all the decrements, other than α, combined. It should be noted, however, that when this is done we are assuming in effect that the decrements are distributed uniformly over the year of age in the single-decrement table α and in the multiple-decrement table β, γ, δ, ...

When an expression for q_x^α in terms of $(am)_x^\alpha$, $(am)_x^\beta$, ... is required, we have, on the assumption that decrements in the single-decrement table α are uniformly distributed over the year of age,

$$q_x^\alpha = \frac{m_x^\alpha}{1+\frac{1}{2}m_x^\alpha}.$$

From formula (20·13), $m_x^\alpha \doteqdot (am)_x^\alpha$, whence

$$q_x^\alpha \doteqdot \frac{(am)_x^\alpha}{1+\frac{1}{2}(am)_x^\alpha}. \tag{20·22}$$

An alternative expression is obtained from formula (20·20) by replacing $(aq)_x^\alpha$ and $(aq)_x$ by the expressions in terms of $(am)_x^\alpha$ and $(am)_x$ developed in the previous chapter on the assumption that the decrements in the *multiple*-decrement table are distributed uniformly over the year of age. This gives

$$q_x^\alpha \doteqdot (am)_x^\alpha\left\{1-\frac{\frac{1}{2}(am)_x^\alpha}{1+\frac{1}{2}(am)_x}\right\} \tag{20·23}$$

or, when only two decrements are involved,

$$q_x^\alpha \doteqdot \frac{(am)_x^\alpha\{1+\frac{1}{2}(am)_x^\beta\}}{1+\frac{1}{2}(am)_x}. \tag{20·24}$$

*20·13. Relationships between $(aq)_x^\alpha$ and q_x^α based on other assumptions as to the distribution of the decrements

It is clear from the two preceding sections that the relationship between $(aq)_x^\alpha$ and q_x^α will depend on the manner in which the decrements which take place during the year of age x to $x+1$ are distributed over that year. Convenient assumptions are

(i) that ${}_tp_x^\alpha \mu_{x+t}^\alpha$ is constant when $0 \leqslant t < 1$, i.e. that the decrements in each *single*-decrement table are distributed uniformly over the year of age,

(ii) that ${}_t(ap)_x\,(a\mu)^\alpha_{x+t}$ is constant when $0 \leqslant t < 1$, i.e. that each decrement in the *multiple*-decrement table is distributed uniformly over the year of age,

(iii) that $(a\mu)^\alpha_{x+t}\,(\equiv \mu^\alpha_{x+t})$ is constant when $0 \leqslant t < 1$.

The first of these assumptions was used as a basis for the formulae developed in sections 11 and 12 of this chapter; we shall now consider assumptions (ii) and (iii).

On assumption (ii)

$$\frac{(aq)^\alpha_x}{(aq)_x} = \frac{\int_0^1 {}_t(ap)_x\,(a\mu)^\alpha_{x+t}\,dt}{\int_0^1 {}_t(ap)_x\,(a\mu)_{x+t}\,dt}$$

$$= \frac{{}_t(ap)_x\,(a\mu)^\alpha_{x+t}}{{}_t(ap)_x\,(a\mu)_{x+t}} = \frac{(a\mu)^\alpha_{x+t}}{(a\mu)_{x+t}}.$$

On assumption (iii)

$$\frac{(aq)^\alpha_x}{(aq)_x} = \frac{\int_0^1 {}_t(ap)_x\,(a\mu)^\alpha_{x+t}\,dt}{\int_0^1 {}_t(ap)_x\,(a\mu)_{x+t}\,dt}$$

$$= \frac{(a\mu)^\alpha_{x+t}\int_0^1 {}_t(ap)_x\,dt}{(a\mu)_{x+t}\int_0^1 {}_t(ap)_x\,dt} = \frac{(a\mu)^\alpha_{x+t}}{(a\mu)_{x+t}}.$$

On either assumption therefore

$$\frac{(aq)^\alpha_x}{(aq)_x} = \frac{(a\mu)^\alpha_{x+t}}{(a\mu)_{x+t}} \equiv \frac{\mu^\alpha_{x+t}}{(a\mu)_{x+t}}.$$

$$\therefore \quad (aq)^\alpha_x\,(a\mu)_{x+t} = (aq)_x\,\mu^\alpha_{x+t}.$$

$$\therefore \quad (aq)^\alpha_x\left\{-\int_0^1 (a\mu)_{x+t}\,dt\right\} = (aq)_x\left\{-\int_0^1 \mu^\alpha_{x+t}\,dt\right\}.$$

$$\therefore \quad (aq)^\alpha_x \log\{1-(aq)_x\} = (aq)_x \log(1-q^\alpha_x).$$

$$\therefore \quad (aq)^\alpha_x = (aq)_x \frac{\log(1-q^\alpha_x)}{\log\{1-(aq)_x\}} \qquad (20{\cdot}25)$$

and

$$\log(1-q^\alpha_x) = \frac{(aq)^\alpha_x}{(aq)_x}\log\{1-(aq)_x\}. \qquad (20{\cdot}26)$$

Formula (20·25) gives $(aq)^\alpha_x$ in terms of q^α_x, q^β_x, etc., because if these are known $(aq)_x$ can be obtained from formula (20·9); formula

(20·26) gives q_x^α in terms of $(aq)_x^\alpha$, $(aq)_x^\beta$, etc. However, neither is in a convenient form, either for application or for comparison with the formulae deduced on assumption (i). It is therefore of some interest to see whether, when two decrements only are involved, the formulae can be converted into others which are more suitable for these purposes.

Expanding both sides of formula (20·26) by the logarithmic series and simplifying the development by writing q_α for q_x^α, α for $(aq)_x^\alpha$ and β for $(aq)_x^\beta$, we have

$$q_\alpha+\frac{q_\alpha^2}{2}+\frac{q_\alpha^3}{3}+\frac{q_\alpha^4}{4}+\ldots=\alpha\left\{1+\frac{\alpha+\beta}{2}+\frac{(\alpha+\beta)^2}{3}+\frac{(\alpha+\beta)^3}{4}+\ldots\right\}. \tag{20·27}$$

By replacing $\alpha+\beta$ on the right-hand side of formula (20·27) by $q_\alpha+q_\beta-q_\alpha q_\beta$, expressing α in the form*

$$\alpha=q_\alpha\{1+aq_\beta+(bq_\alpha q_\beta+cq_\beta^2)+(dq_\alpha^2q_\beta+eq_\alpha q_\beta^2+fq_\beta^3)+\ldots\}$$

and equating coefficients, the following formula giving $(aq)_x^\alpha$ in terms of q_x^α and q_x^β may be obtained:

$$(aq)_x^\alpha=q_\alpha\{1-\tfrac{1}{2}q_\beta+\tfrac{1}{12}(q_\alpha q_\beta-q_\beta^2)+\tfrac{1}{24}(q_\alpha^2q_\beta-q_\beta^3)+\ldots\}. \tag{20·28}$$

Similarly, expressing q_α on the left-hand side of formula (20·27) in the form

$$q_\alpha=\alpha\{1+a\beta+(b\alpha\beta+c\beta^2)+(d\alpha^2\beta+e\alpha\beta^2+f\beta^3)+\ldots\}$$

and equating coefficients, we have the following formula giving q_x^α in terms of $(aq)_x^\alpha$ and $(aq)_x^\beta$:

$$q_x^\alpha=\alpha\{1+\tfrac{1}{2}\beta+\tfrac{1}{6}(\alpha\beta+2\beta^2)+\tfrac{1}{24}(2\alpha^2\beta+7\alpha\beta^2+6\beta^3)+\ldots\}. \tag{20·29}$$

Comparison of formula (20·28) with formula (20·12), and of formula (20·29) with formula (20·21), shows that the effect of making assumption (ii) or (iii) instead of assumption (i) is very small. Formulae (20·28) and (20·29) do not lead to any new formulae that are convenient in practice, but they show that the formulae deduced in the preceding section of this chapter, although based

* Since, when $q_\alpha=0$, $\alpha=0$, it follows that the expansion of α in terms of q_α and q_β can contain no terms in q_β, q_β^2, q_β^3, etc. Also, when $q_\beta=0$, $q_\alpha=\alpha$, whence the term of the first degree in the expansion must be q_α and there can be no terms in q_α^2, q_α^3, etc.

on assumption (i), can be regarded for practical purposes as being equally suitable on assumption (ii) or (iii). It will be appreciated that one is rarely in the position of being able to say which of the three assumptions is most nearly in accordance with the facts.

It is of interest to consider the form of the relationships between $(aq)_x^\alpha$ and q_x^α, when only two decrements are involved, without introducing any assumptions as to the distribution of the decrements over the year of age.

From formula (20·10)

$$q_\alpha + q_\beta - q_\alpha q_\beta = \alpha + \beta.$$

Expressing α in the form

$$\alpha = q_\alpha\{1 + aq_\beta + (bq_\alpha q_\beta + cq_\beta^2) + (dq_\alpha^2 q_\beta + eq_\alpha q_\beta^2 + fq_\beta^3) + \ldots\}$$

and expressing β similarly, we have

$$q_\alpha + q_\beta - q_\alpha q_\beta = q_\alpha + q_\beta + 2aq_\alpha q_\beta + (b+c)(q_\alpha^2 q_\beta + q_\alpha q_\beta^2) + (d+f)(q_\alpha^3 q_\beta + q_\alpha q_\beta^3) + 2eq_\alpha^2 q_\beta^2 + \ldots.$$

Equating coefficients we find that $a = -\frac{1}{2}$, $c = -b$, $f = -d$, $e = 0$, ... Hence, writing ϵ_1 for b and ϵ_2 for d, we have

$$(aq)_x^\alpha = q_\alpha\{1 - \tfrac{1}{2}q_\beta + \epsilon_1 q_\alpha q_\beta - \epsilon_1 q_\beta^2 + \epsilon_2 q_\alpha^2 q_\beta - \epsilon_2 q_\beta^3 + \ldots\}. \quad (20{\cdot}30)$$

By a similar method it can be shown that

$$q_x^\alpha = \alpha\{1 + \tfrac{1}{2}\beta + (\tfrac{1}{4} - \epsilon_1)\alpha\beta + (\tfrac{1}{4} + \epsilon_1)\beta^2 + (\tfrac{1}{8} - \epsilon_2)\alpha^2\beta + (\tfrac{3}{8} - \epsilon_1)\alpha\beta^2 + (\tfrac{1}{8} + \epsilon_1 + \epsilon_2)\beta^3 + \ldots\}. \quad (20{\cdot}31)$$

These two formulae are of theoretical interest but in the absence of some means of ascertaining the values of ϵ_1 and ϵ_2 they are of little practical use. By comparing formula (20·12) with formula (20·30), and formula (20·21) with formula (20·31), we see that assumption (i) is equivalent to the assumption that $\epsilon_1 = 0$ and $\epsilon_2 = 0$. By comparing formula (20·28) with formula (20·30), and formula (20·29) with formula (20·31), we see that assumption (ii) or (iii) is equivalent to the assumption that $\epsilon_1 = \frac{1}{12}$ and $\epsilon_2 = \frac{1}{24}$.

It is interesting to notice that if $q_\alpha = q_\beta$ the terms involving ϵ_1 and ϵ_2 in formula (20·30) cancel out; it follows that if $q_x^\alpha \doteq q_x^\beta$ formula (20·12), which is correct on assumption (i), will be approximately correct on any other assumption as to the distribution of the decrements over the year of age. Similarly, if $\alpha=\beta$, the terms involving ϵ_1 and ϵ_2 in formula (20·31) cancel out; it follows that if $(aq)_x^\alpha \doteq (aq)_x^\beta$ formula (20·20), which is very nearly exact on assumption (i), will be approximately correct on any other assumption.

20·14. Abnormal incidence of decrements

Hitherto it has been assumed that all decrements occur continuously. When this is not the case, appropriate formulae may be deduced from first principles.

As an illustration, let us consider a table with two decrements, α which is distributed uniformly over the year of age x to $x+1$ and β which is concentrated at age $x+k$, where $0 \leqslant k < 1$.

$$(aq)_x^\alpha = \int_0^1 {}_t(ap)_x \, (a\mu)_{x+t}^\alpha \, dt$$

$$= \int_0^1 {}_tp_x^\alpha \cdot {}_tp_x^\beta \mu_{x+t}^\alpha \, dt.$$

Now $\qquad {}_tp_x^\beta = 1, \quad 0 \leqslant t < k,$

$$ {}_tp_x^\beta = 1 - q_x^\beta, \quad k \leqslant t < 1,$$

and $\qquad {}_tp_x^\alpha \mu_{x+t}^\alpha = q_x^\alpha, \quad 0 \leqslant t < 1.$

$$\therefore \quad (aq)_x^\alpha = q_x^\alpha\{k + (1-k)(1-q_x^\beta)\}$$

$$= q_x^\alpha\{1-(1-k)\,q_x^\beta\}.$$

Also since $\quad (aq)_x^\alpha + (aq)_x^\beta = q_x^\alpha + q_x^\beta - q_x^\alpha q_x^\beta,$

$$(aq)_x^\beta = q_x^\beta(1 - kq_x^\alpha).$$

If $k=0$, and
$$\left.\begin{aligned}(aq)_x^\alpha &= q_x^\alpha(1-q_x^\beta)\\ (aq)_x^\beta &= q_x^\beta.\end{aligned}\right\} \qquad (20·32)$$

If $k=\tfrac{1}{2}$, and
$$\left.\begin{aligned}(aq)_x^\alpha &= q_x^\alpha(1-\tfrac{1}{2}q_x^\beta)\\ (aq)_x^\beta &= q_x^\beta(1-\tfrac{1}{2}q_x^\alpha).\end{aligned}\right\} \qquad (20·33)$$

If $k=1$, and
$$\left.\begin{aligned}(aq)_x^\alpha &= q_x^\alpha\\ (aq)_x^\beta &= q_x^\beta(1-q_x^\alpha).\end{aligned}\right\} \qquad (20·34)$$

It will be seen that formulae (20·33) are identical with formulae (20·12) which were developed on the assumption that the decrements in both single-decrement tables are distributed uniformly over the year. It will also be seen that formulae (20·32) and (20·34) could have been derived without the assumption that the α-decrements are distributed *uniformly* over the year; these formulae are in agreement with formulae (20·2), (20·3), (20·4) and (20·5).

More complex examples involving three or more decrements can be worked out. In a case such as that mentioned in section 4 of Chapter 19, it is convenient to regard the retirements which take place at exact age 60 and those which are distributed over the year of age 60 to 61 as two separate decrements.

EXAMPLES

Example 20·1

(Tables for use in solving this and other examples will be found at the end of this volume.)

Among the employees of a certain firm, retirement may take place after age 57, but is compulsory at age 60.

When an employee enters the service, the firm sets aside a sum sufficient to provide for the payment of £500 on his death while in the service, calculated on the following basis:

Mortality	Hypothetical Select Table, Ultimate mortality
Interest	2 per cent
Withdrawals	Nil
Retirements	As set out below:

Age	Central rate of retirement
57	·25
58	·05
59	·05

What sum must be set aside for an entrant aged 40?

Solution

From ages 40 to 56 inclusive, we may use the Hypothetical Select Table values of l_x and d_x as $(al)_x$ and $(ad)_x$ of the double-decrement table. Also, $(al)_{57}=l_{57}$. In order to obtain $(ad)_x^d$ and $(ad)_x^r$ at ages 57 to 59, we have from formula (20·12) or (20·16)

$$(aq)_x^d=\frac{q_x^d}{1+\frac{1}{2}m_x^r}$$

and from formula (20·16)

$$(aq)_x^r=\frac{m_x^r(1-\frac{1}{2}q_x^d)}{1+\frac{1}{2}m_x^r}.$$

These relationships are exact on the assumption that the decrements in the single-decrement tables are distributed uniformly over the year of age.

We are not told whether the central rates given in the question are m_x^r or $(am)_x^r$ but the difference between the two is negligible. We may therefore proceed as follows:

x	q_x^d	m_x^r	$1+\frac{1}{2}m_x^r$	$1-\frac{1}{2}q_x^d$	$(aq)_x^d$	$(aq)_x^r$
57	·01432	·25	1·125	·99284	·01273	·22063
58	·01588	·05	1·025	·99206	·01549	·04839
59	·01759	·05	1·025	·99120	·01716	·04835

x	$(al)_x$	$(ad)_x^d$	$(ad)_x^r$
57	424,215	5,400	93,595
58	325,220	5,038	15,737
59	304,445	5,224	

Therefore, the sum to be set aside by the firm is

$$500\times 1{\cdot}01\times\frac{M_{40}-M_{57}+v^{58}(ad)_{57}^d+v^{59}(ad)_{58}^d+v^{60}(ad)_{59}^d}{D_{40}}$$

$$=505\,\frac{18{,}459+1{,}712+1{,}566+1{,}592}{214{,}984}$$

$$=505\,\frac{23{,}329}{214{,}984}=54{\cdot}800=\pounds 54\ 16s.$$

Rough check

$500\times 1{\cdot}01\times\dfrac{M_{40}-M_{60}}{D_{40}}=\pounds 58$ approx. The answer should be less than this.

Example 20·2

In a certain double-decrement table the force of mortality is constant at ·01 and the force of withdrawal is constant at ·1. Assuming that the decrements are not selective, calculate accurately the independent and the dependent rates of mortality and withdrawal.

Test the arithmetical accuracy of formulae (20·12), (20·18), (20·20) and (20·29).

Solution

$$q_x^d = 1 - \exp\left(-\int_0^1 \cdot 01\, dt\right)$$
$$= 1 - \exp(-\cdot 01)$$
$$= 1 - \cdot 990050$$
$$= \cdot 009950.$$

Similarly,
$$q_x^w = 1 - \cdot 904837$$
$$= \cdot 095163.$$

Now
$$(a\mu)_{x+t}^d \equiv \mu_{x+t}^d = \cdot 01,$$
$$(a\mu)_{x+t}^w = \cdot 1$$
and
$$(a\mu)_{x+t} = \cdot 11.$$

Therefore
$$(aq)_x = 1 - \exp\left(-\int_0^1 \cdot 11\, dt\right)$$
$$= 1 - \cdot 895834$$
$$= \cdot 104166,$$
$$(aq)_x^d = \frac{\cdot 01}{\cdot 11} \times \cdot 104166 = \cdot 009470$$
and
$$(aq)_x^w = \frac{\cdot 1}{\cdot 11} \times \cdot 104166 = \cdot 094696.$$

Using the above values of q_x^d and q_x^w in formula (20·12), we have

$$(aq)_x^d = \cdot 009477 \qquad \text{(True } \cdot 009470)$$
and
$$(aq)_x^w = \cdot 094690 \qquad \text{(True } \cdot 094696).$$

Using the values of $(aq)_x^d$ and $(aq)_x^w$ in formula (20·18), we have

$$q_x^d = \cdot 009941 \qquad (\text{True } \cdot 009950)$$

and

$$q_x^w = \cdot 095147 \qquad (\text{True } \cdot 095163),$$

in formula (20·20),

$$q_x^d = \cdot 009943 \qquad (\text{True } \cdot 009950)$$

and

$$q_x^w = \cdot 095169 \qquad (\text{True } \cdot 095163),$$

and in formula (20·29),

$$q_x^d = \cdot 009950 \qquad (\text{True } \cdot 009950)$$

and

$$q_x^w = \cdot 095162 \qquad (\text{True } \cdot 095163).$$

Note

It will be seen that in this particular case formula (20·29) is the most accurate. This was to be expected, since one of the alternative assumptions on which this formula is based is that the forces are constant over the year of age.

Example 20·3

In a certain pension fund, it is expected that the independent rates of mortality will be those of the English Life Table No. 10, Males. An investigation of the experience of the fund shows a dependent rate of retirement of ·2 for the year of age $59 < x \leqslant 60$. 90 per cent of the retirements in this age range occur at exact age 60. There are no withdrawals during this year of age.

On the basis of these figures, develop as accurately as possible the portion of the multiple-decrement table for the year of age $59 < x \leqslant 60$.

Solution

$$q_{59}^d = \cdot 02224$$

$$(aq)_{59}^r = \cdot 2$$

Out of 100,000 lives aged 59, it is expected that 2,000 will retire over the age range $59 < x < 60$ and 18,000 at exact age 60. We will use r' to represent retirements over the year of age, and r'' to represent retirements at exact age 60. Then

$$(aq)_{59}^{r'} = \cdot 02,$$

$$\begin{aligned}(aq)_{59}^{d} &= q_{59}^{d}(1 - \tfrac{1}{2}q_{59}^{r'}) \\ &= q_{59}^{d}\left(1 - \frac{1}{2} \cdot \frac{(aq)_{59}^{r'}}{1 - \frac{1}{2}q_{59}^{d}}\right) \\ &= \cdot 02224\left(1 - \frac{1}{2} \cdot \frac{\cdot 02}{1 - \frac{1}{2} \times \cdot 02224}\right) \\ &= \cdot 02202.\end{aligned}$$

Hence, if $(al)_{59} = 100{,}000,$

it follows that

$$(ad)_{59}^{d} = 2{,}202,$$

$$(ad)_{59}^{r'} = 2{,}000,$$

$$(ad)_{59}^{r''} = 18{,}000 \text{ (retirements at exact age 60)},$$

$$(ad)_{59}^{r} = 20{,}000$$

and $(al)_{60} = 77{,}798.$

CHAPTER 21

THE JOINT-LIFE MORTALITY TABLE: SOME STANDARD DEVIATIONS

21·1. The joint-life mortality table

We have seen in Volume I that the probability that a life (x), taken at random from an indefinitely large number of lives who are assumed to experience, as a whole, the mortality of the mortality table, will survive n years is ${}_np_x$. If two such lives, (x) and (y), are each taken at random and if the probabilities of their survival are independent, the probability that both lives will survive n years is ${}_np_x \times {}_np_y$, which is denoted by the symbol ${}_np_{xy}$. We therefore have the relationship

$$ {}_np_{xy} = {}_np_x \times {}_np_y. \tag{21·1}$$

Now, since
$$ {}_np_x = \exp\left\{-\int_0^n \mu_{x+t}\,dt\right\},$$
it follows that
$$ {}_np_{xy} = \exp\left\{-\int_0^n \mu_{x+t}\,dt\right\} \times \exp\left\{-\int_0^n \mu_{y+t}\,dt\right\}$$
$$ = \exp\left\{-\int_0^n (\mu_{x+t} + \mu_{y+t})\,dt\right\}.$$

If, therefore, we write
$$\mu_{xy} = \mu_x + \mu_y,$$
we have
$$ {}_np_{xy} = \exp\left\{-\int_0^n \mu_{x+t:y+t}\,dt\right\}. \tag{21·2}$$

If we also write
$$l_{xy} = l_x \times l_y,$$
we have
$$ {}_np_{xy} = \frac{l_{x+n:y+n}}{l_{xy}} \tag{21·3}$$
and
$$ p_{xy} = \frac{l_{x+1:y+1}}{l_{xy}}. \tag{21·4}$$

Taking logarithms of the expressions on the right-hand sides of formulae (21·2) and (21·3) and equating them, we have

$$\log_e l_{x+n:y+n} - \log_e l_{xy} = -\int_0^n \mu_{x+t:y+t}\,dt,$$

whence $$\frac{d}{dt}\log_e l_{x+t:y+t} = -\mu_{x+t:y+t}$$

or $$\frac{1}{l_{x+t:y+t}}\frac{d}{dt}l_{x+t.y+t} = -\mu_{x+t:y+t}. \tag{21·5}$$

Multiplying both sides of formula (21·5) by $l_{x+t:y+t}$ and integrating, we have

$$l_{xy} = \int_0^\infty l_{x+t:y+t}\mu_{x+t:y+t}dt. \tag{21·6}$$

If ${}_nq_{xy}$ denotes the probability that at least one of the two lives (x) and (y) will die within n years,

$$\left.\begin{aligned} {}_nq_{xy} &= 1 - {}_np_{xy} \\ &= \frac{l_{xy} - l_{x+n:y+n}}{l_{xy}} \end{aligned}\right\} \tag{21·7}$$

and $$q_{xy} = 1 - p_{xy} = \frac{d_{xy}}{l_{xy}}, \tag{21·8}$$

where $$d_{xy} = l_{xy} - l_{x+1:y+1}. \tag{21·9}$$

This definition of d_{xy} should be noted.

From formulae (21·6) and (21·9)

$$d_{xy} = \int_0^1 l_{x+t:y+t}\mu_{x+t:y+t}dt, \tag{21·10}$$

whence $$q_{xy} = \int_0^1 {}_tp_{xy}\mu_{x+t:y+t}dt. \tag{21·11}$$

From formula (21·1)

$$1 - q_{xy} = (1 - q_x)(1 - q_y),$$

whence $$q_{xy} = q_x + q_y - q_x \times q_y. \tag{21·12}$$

The function l_{xy} and its related joint-life functions, d_{xy}, p_{xy}, etc., constitute the joint-life mortality table.

The main purpose of the joint-life mortality table is to facilitate the calculations that are frequently required in connexion with assurances and annuities effected on two lives. Occasionally, however, functions in respect of more than two lives are required. The extension of the formulae developed above to three or more lives will be sufficiently obvious.

It is not essential that both (x) and (y) should be regarded as taken at random from the same group of lives. It frequently happens that (x) and (y) are regarded as taken at random from two groups of lives who are assumed to experience respectively the mortality of two tables, say Table I and Table II. For example, (x) may be an 'assured' life and (y) an 'annuitant' life or (x) may be a male and (y) a female. In these circumstances, formulae (21·1) to (21·12) will still apply, provided that all single-life functions relating to (x) are taken from Table I and that those relating to (y) are taken from Table II. Thus we shall have

$$l^{\mathrm{I}\ \mathrm{II}}_{x:y} = l^{\mathrm{I}}_x \times l^{\mathrm{II}}_y$$

or (if (x) is a male and (y) a female)

$$l^{\mathrm{MF}}_{x:y} = l^{\mathrm{M}}_x \times l^{\mathrm{F}}_y$$

and so on. Where no ambiguity can arise, the affixes I and II (or M and F) may be omitted. In theoretical work, the association of the letter x with Table I and the letter y with Table II will usually be sufficient to prevent ambiguity; the avoidance of ambiguity in practical work, where the ages of the lives are represented by figures and not letters, will be considered in section 5 of this chapter.

21·2. Selection

Formulae (21·1) to (21·12) may easily be extended to select tables. When both lives are select it is usual to write $[xy]$ instead of $[x]:[y]$. Thus

$$l_{[xy]} = l_{[x]} \times l_{[y]}$$

and so on.

Sometimes a select life is combined with one who is no longer select. In such cases functions like $l_{[x]:[y]+t}$ will occur. This does not introduce any theoretical difficulties.

21·3. The joint-life mortality table: an alternative approach

We saw in section 5 of Chapter 19 that the relationships between the total-decrement functions of the multiple-decrement table are exactly the same as those between the corresponding functions of the single-life mortality table. We now see that there is an even closer correspondence between the formulae relating to the joint-life table and those relating to the multiple-decrement table. Formulae (21·2) to (21·11) correspond to formulae which have been developed in Chapter 19 in respect of the multiple-decrement table; formulae (21·1) and (21·12) correspond to formulae in Chapter 20 connecting the multiple-decrement table with its related single-decrement tables. In fact, we can regard the joint existence of two lives (x) and (y) as being a status subject to two modes of decrement, the death of (x) and the death of (y); it follows that the joint-life mortality table may be regarded as a double-decrement table. We could therefore have obtained formulae (21·1) to (21·12) from the corresponding formulae in Chapters 19 and 20 by merely altering the notation in accordance with Table 21·1.

TABLE 21·1

Function in the double-decrement table or its related single-decrement table	Corresponding function in the joint-life mortality table or its related single-life mortality table
$(al)_x$	l_{xy}
$(a\mu)_x$	μ_{xy}
${}_n(ap)_x$	${}_np_{xy}$
$(ap)_x$	p_{xy}
${}_n(aq)_x$	${}_nq_{xy}$
$(aq)_x$	q_{xy}
$(ad)_x$	d_{xy}
${}_np_x^\alpha$	${}_np_x$
${}_np_x^\beta$	${}_np_y$
q_x^α	q_x
q_x^β	q_y

Since (x) and (y) are assumed to be taken at random independently of each other, the joint-life mortality table, if regarded as a double-decrement table, presents the unusual feature of having decrements which are non-selective.

The resemblance between the joint-life mortality table and the multiple-decrement table, although of some theoretical interest, is of little or no practical value, since problems involving two lives are more conveniently dealt with by methods which will be demonstrated in later chapters of this book than by the technique of the multiple-decrement table.

21·4. Calculation of numerical values

Although tables of commutation functions based on the joint-life mortality table are frequently constructed in order to facilitate the evaluation of joint-life annuities and assurances, tables of numerical values of the non-monetary functions l_{xy}, p_{xy}, etc., are rarely constructed, since individual values can easily be obtained from the single-life mortality table or tables if they are needed.

Should a column of numerical values of l_{xy}, $l_{x+1:y+1}$, etc., be required, it is most conveniently constructed by multiplying l_x by l_y, l_{x+1} by l_{y+1}, and so on, rather than by starting with the radix $l_{xy}=l_x \times l_y$ and multiplying successively by p_{xy}, $p_{x+1:y+1}$, etc.

21·5. The order of the ages in the suffix

When functions in respect of both (x) and (y) are taken from the same mortality table, the order of the ages in the suffix of a non-monetary joint-life function is immaterial. Thus l_{xy} is identical with l_{yx} and $l_{30:40}$, for example, is identical with $l_{40:30}$. However, as will be seen in later chapters of this Volume, it may be advantageous, when numerical values of commutation functions are being calculated, to adopt the convention that the life whose age is mentioned first in the suffix of the commutation symbol is (x). It makes for neatness and consistency to extend this convention to the non-monetary functions.

When functions in respect of (x) and (y) are taken from two different mortality tables (say Table I and Table II respectively) the convention that x shall always come before y in the suffix of non-monetary as well as monetary functions has the further advantage that it helps to avoid ambiguity in practical work. If the symbols l_{xy} and l_{yx} were used indiscriminately, the symbol $l_{30:40}$ could mean either $l^{\mathrm{I}}_{30}{}^{\mathrm{II}}_{:40}$ or $l^{\mathrm{II}}_{30}{}^{\mathrm{I}}_{:40}$ and the affixes I and II would have to be added in order to indicate which was meant; if, however, it is understood that the age of (x) (the Table I life) is always placed before the age of (y) (the Table II life), the symbol $l_{30:40}$ can have only one meaning.

21·6. The independence of the probabilities

Formulae (21·1) to (21·12) have been built up on the assumption that the probabilities ${}_np_x$ and ${}_np_y$ are independent. In practice joint-life functions are required mainly in connexion with annuities and assurances on lives who are connected with each other in some way. For example, the lives may be partners in business or husband and wife. In such circumstances the assumption that both lives are taken at random is, strictly speaking, not correct, and the probabilities ${}_np_x$ and ${}_np_y$ are therefore not independent. Thus, if (x) is killed in a motor accident, (y) is much more likely to be killed in the same accident if he or she is (x)'s business partner or wife than if (x) and (y) have both been taken at random. Again, it is well-known that the death of a husband or wife sometimes appears to accelerate the death of the surviving spouse. However, in the calculation of net premiums and reserves, it is usual to proceed as if the probabilities were independent; the fact that this procedure may not be correct should be borne in mind when the basis for office premiums is being fixed.

Similar remarks apply to other forms of annuity and assurance involving two or more lives, such as reversionary annuities and contingent assurances, which are considered later in this book.

21·7. Some standard deviations

The remainder of this chapter will be devoted to obtaining expressions for certain standard deviations. These problems do not require the use of the theory of the joint-life mortality table but they imply a consideration of the joint survival of a number of lives and it is therefore appropriate to consider them here.

21·8. The standard deviation of the number of deaths in a group of *n* lives

Let us consider a group of n lives aged x, each of whom is taken at random from an indefinitely large number of lives who are assumed to experience, as a whole, the mortality of the single-life mortality table.

The expected number of survivors to age $x+1$ is np_x and the expected number of deaths between ages x and $x+1$ is nq_x. The probability that exactly r lives will die before attaining age $x+1$ is

$$\binom{n}{r} p_x^{n-r} q_x^r.$$

Since we are concerned with a theoretical model we can imagine a very large number of groups similarly taken at random. The number of deaths will vary from group to group but the mean number of deaths among the groups will be the same as the expected number for one group, i.e. nq_x, and the expected relative frequency of groups with r deaths will be

$$\binom{n}{r} p_x^{n-r} q_x^r.$$

The moments of the binomial distribution are given in most textbooks on statistics (see for example *Statistics* by Johnson and Tetley, Vol. I, p. 195, where p and q should be interchanged for the present purpose). The first two moments are

$$\mu_1' = nq$$

and

$$\mu_2' = nq + n(n-1)\,q^2.$$

$$\therefore \quad \mu_2 = nq - nq^2 = npq.$$

It follows that the expected number of deaths is nq_x (as has been mentioned above) and that the standard deviation of the number of deaths among the groups is

$$\sigma = \sqrt{\mu_2} = \sqrt{(np_x q_x)}. \tag{21·13}$$

In applying formula (21·13) to an actual experience it is important to keep clearly in mind the underlying assumption that each unit* in the investigation is taken at random independently of all the others. This assumption may not always be fulfilled. A discussion of the point would be beyond the scope of this book.

* In an investigation into the mortality of assured lives or annuitants, the unit may be a policy and not a life.

*21·9. The standard deviation of the expectation of life

From section 3 of Chapter 14 (Volume I of this book) the first moment of the curve of deaths is

$$\mu_1' = \frac{1}{l_x}\int_0^\infty t\, l_{x+t}\mu_{x+t}\,dt$$

$$= \frac{1}{l_x}\int_0^\infty l_{x+t}\,dt = \frac{T_x}{l_x} = \mathring{e}_x$$

and the second moment is

$$\mu_2' = \frac{1}{l_x}\int_0^\infty t^2 l_{x+t}\mu_{x+t}\,dt$$

$$= \frac{2}{l_x}\int_0^\infty t\, l_{x+t}\,dt = 2\frac{Y_x}{l_x},$$

where $$Y_x = \int_0^\infty t\, l_{x+t}\,dt = \int_0^\infty T_{x+t}\,dt.$$

Hence the standard deviation, σ, of the expectation of life is given by

$$\sigma^2 = \mu_2' - \mu_1'^2$$

$$= 2\frac{Y_x}{l_x} - \mathring{e}_x^2. \qquad (21\cdot 14)$$

*21·10. The standard deviation of the profit or loss on a policy

Consider an n-year endowment assurance of 1 by annual premiums on (x), the sum assured being payable at the end of the year of death. The value at the date of entry into assurance of the loss if a claim occurs in the tth year is the present value of the sum assured, v^t, less the present value of an annuity-certain of the premiums, $P_{[x]:\overline{n}|}\ddot{a}_{\overline{t}|}$, i.e.

$$v^t - P_{[x]:\overline{n}|}\ddot{a}_{\overline{t}|} = v^t - \frac{A_{[x]:\overline{n}|}}{d\ddot{a}_{[x]:\overline{n}|}}(1 - v^t)$$

$$= \frac{1}{d\ddot{a}_{[x]:\overline{n}|}}(v^t - A_{[x]:\overline{n}|}) = L \quad \text{(say)}.$$

The mean loss

$$=\sum_{t=1}^{n} L_t \frac{d'_{[x]+t-1}}{l_{[x]}} \quad \text{(where } d'_{[x]+t-1}=d_{[x]+t-1}, \qquad t<n,$$

$$d'_{[x]+t-1}=d_{[x]+n-1}+l_{[x]+n},\ t=n)$$

$$=\frac{1}{d\ddot{a}_{[x]:\overline{n}|}}(A_{[x]:\overline{n}|}-A_{[x]:\overline{n}|})$$

$$=0 \quad \text{as was to be expected.}$$

The square of the standard deviation of the profit or loss is

$$\sum_{t=1}^{n} L_t^2 \frac{d'_{[x]+t-1}}{l_{[x]}}=\frac{1}{d^2(\ddot{a}_{[x]:\overline{n}|})^2}\cdot\frac{1}{l_{[x]}}\sum_{t=1}^{n} d'_{[x]+t-1}\{v^{2t}-2v^t A_{[x]:\overline{n}|}+(A_{[x]:\overline{n}|})^2\}$$

$$=\frac{1}{d^2(\ddot{a}_{[x]:\overline{n}|})^2}\{A'_{[x]:\overline{n}|}-2(A_{[x]:\overline{n}|})^2+(A_{[x]:\overline{n}|})^2\},$$

where $A'_{[x]:\overline{n}|}$ is calculated at a rate of interest i' such that $v'=v^2$. Hence the standard deviation, σ, is given by

$$\sigma^2=\frac{A'_{[x]:\overline{n}|}-(A_{[x]:\overline{n}|})^2}{d^2(\ddot{a}_{[x]:\overline{n}|})^2}. \qquad (21\cdot15)$$

This formula is employed in the 'individual theory of risk' which has been developed extensively on the continent of Europe. It is not proposed to discuss the theory of risk in this book.

*21·11. The standard deviation of the rate of sickness

Let us assume that a hypothetical group of l_x lives suffers sickness of duration a to $a+b$ in the year of age x to $x+1$ according to the following distribution:

Number who have no sickness (of this duration) during the year $=l_x \pi^{a/b}_{x:0}$,

Number whose sickness (of this duration) during the year lasts for time $t=l_x \pi^{a/b}_{x:t}$.

Then
$$\pi^{a/b}_{x:0}+\int_0^b \pi^{a/b}_{x:t}\,dt=1.$$

The first moment of the sickness distribution is

$$0\times \pi^{a/b}_{x:0}+\int_0^b t\pi^{a/b}_{x:t}\,dt.$$

This is also the expected period of sickness (of duration a to $a+b$) of one life taken at random. It may therefore be referred to as 'the rate of sickness of duration a to $a+b$' and denoted by the symbol $s_x^{a/b}$. (It should be noted that this rate of sickness is based on a year as a unit of time, whereas the rate of sickness mentioned in section 8 of Chapter 17 is measured in weekly units.) The second moment of the sickness distribution is

$$\int_0^b t^2 \pi_{x:t}^{a/b} dt$$

and the standard deviation, σ, of the rate of sickness is given by

$$\sigma^2 = \int_0^b t^2 \pi_{x:t}^{a/b} dt - (s_x^{a/b})^2. \tag{21·16}$$

This is the standard deviation in respect of one life taken at random. For n lives the expected total sickness is n times the sickness rate and the square of the standard deviation of the expected total sickness is n times the expression given above.

EXAMPLES

Example 21·1

Express in terms of the single-life probabilities p_x and p_y the probability in respect of two lives (x) and (y) that:

(*a*) both lives will survive a year,
(*b*) at least one life will survive a year,
(*c*) exactly one life will survive a year,
(*d*) not more than one life will survive a year,
(*e*) neither life will survive a year,
(*f*) at least one life will die within a year,
(*g*) not more than one life will die within a year.

Solution

(*a*) $p_x p_y = p_{xy}$.

(*b*) $p_x p_y + p_x q_y + p_y q_x$
$= p_x + p_y - p_{xy}$.

(*c*) $p_x q_y + p_y q_x$
$= p_x + p_y - 2p_{xy}$.

(d) $p_x q_y + p_y q_x + q_x q_y$
$= 1 - p_{xy}$.

(e) $q_x q_y = 1 - p_x - p_y + p_{xy}$.

(f) $q_x p_y + q_y p_x + q_x q_y$
$= 1 - p_{xy}$. (This is, of course, the same as (d).)

(g) $1 - q_x q_y$
$= p_x + p_y - p_{xy}$. (This is, of course, the same as (b).)

Example 21·2

The sum of £10,000 is to be divided 30 years hence between a charitable trust and such of four lives now aged 30 as survive, the participants taking equal shares. Find the amount of the institution's expectation according to the ultimate mortality of the Hypothetical Select Table.

Solution

$$\text{Expectation} = 10{,}000\,(\tfrac{1}{5}p^4 + \tfrac{1}{4}.4p^3q + \tfrac{1}{3}.6p^2q^2 + \tfrac{1}{2}.4pq^3 + q^4)$$
$$= 10{,}000\,(\cdot 2p^4 + p^3q + 2p^2q^2 + 2pq^3 + q^4),$$

where $p = {}_{30}p_{30}$ and $q = {}_{30}q_{30}$.

$\log l_{60} = 5{\cdot}60666$
$\log l_{30} = 5{\cdot}68673$
$\therefore \log p = \bar{1}{\cdot}91993$. $\quad \therefore p = {\cdot}8316$ and $q = {\cdot}1684$.
$\therefore \log q = \bar{1}{\cdot}22634$.

$\log p^4 = \bar{1}{\cdot}6797$	$\cdot 4783 \times \cdot 2$	$= \cdot 0957$
$\log p^3 = \bar{1}{\cdot}7598$		
$\log q = \bar{1}{\cdot}2263$		
$\bar{2}{\cdot}9861$	$\cdot 0968$	$= \cdot 0968$
$\log p^2 = \bar{1}{\cdot}8399$		
$\log q^2 = \bar{2}{\cdot}4527$		
$\bar{2}{\cdot}2926$	$\cdot 0196 \times 2$	$= \cdot 0392$
$\log p = \bar{1}{\cdot}9199$		
$\log q^3 = \bar{3}{\cdot}6790$		
$\bar{3}{\cdot}5989$	$\cdot 0040 \times 2$	$= \cdot 0080$
$\log q^4 = \bar{4}{\cdot}9054$	$\cdot 0008$	$= \cdot 0008$
		$\cdot 2405$

The expectation is therefore £2,405.

Example 21·3

(*a*) An office is asked to quote for a group of one-year temporary assurances of £100 each on 100 lives, all aged 56. On the basis of the English Life Table No. 10, Males, and 4 per cent interest, and on the assumption that claims are payable at the end of the period, what is the minimum aggregate premium the office must charge, excluding expenses, in order that it may have at least an even chance of making a profit?

Solution

The probability that there will be no deaths $= p^{100} = \cdot 172$.

The probability of 1 death $= 100p^{99}q \qquad = \cdot 306$.

The probability of 2 deaths $= \dfrac{100 \times 99}{2} p^{98}q^2 \qquad = \cdot 269$.

Therefore the probability of not more than 1 death $= \cdot 478$

and the probability of not more than 2 deaths $= \cdot 747$.

If there are to be two deaths the office must have in hand at the outset a sum of at least $200v =$ £192. 6*s*. 2*d*. in order to be able to pay the claims at the end of the year.

In order to have at least an even chance of making a profit it must therefore charge an aggregate net premium of at least £192. 6*s*. 3*d*.

Note

The normal net premium is $100 \times 100vq = 167 \cdot 7$.

Example 21·4

Develop an expression for the standard deviation of $\bar{a}_x$.

Solution

$$\mu_1' = \frac{1}{l_x}\int_0^\infty \bar{a}_{\overline{t|}}\, l_{x+t}\mu_{x+t}\,dt = \bar{a}_x$$

and

$$\mu_2' = \frac{1}{l_x}\int_0^\infty (\bar{a}_{\overline{t|}})^2\, l_{x+t}\mu_{x+t}\,dt.$$

Now
$$(\bar{a}_{\overline{t}|})^2 = \delta^{-2}(1-v^t)^2$$
$$= \delta^{-2}(1-2v^t+v^{2t})$$
$$= \delta^{-1}\left\{\frac{2(1-v^t)}{\delta} - \frac{2(1-v^{2t})}{2\delta}\right\}$$
$$= \frac{2}{\delta}(\bar{a}_{\overline{t}|} - \bar{a}'_{\overline{t}|}),$$
where
$$v' = v^2.$$

Hence the standard deviation, σ, is given by

$$\sigma^2 = \mu'_2 - \mu'^2_1 = \frac{2}{\delta}(\bar{a}_x - \bar{a}'_x) - \bar{a}^2_x.$$

Note

The student will find it an interesting exercise to verify that, as $\delta \to 0$,

$$\frac{\bar{a}_x - \bar{a}'_x}{\delta} \to \frac{Y_x}{l_x},$$

so reconciling the above answer with formula (21·14).

CHAPTER 22

JOINT-LIFE AND LAST SURVIVOR ANNUITIES, ASSURANCES, ETC.

22·1. Joint-life endowments, annuities and assurances

The values of joint-life endowments, annuities and assurances can be represented by expressions similar to those developed in Volume I for the corresponding single-life functions. It will be sufficient to set out here some of the more important expressions.

To simplify the formulae it will be assumed that only two lives are involved. The extension to three or more lives will be obvious.

The value of a pure endowment of 1, payable if (x) and (y) both survive t years, is

$$ {}_tE_{xy} = v^t\, {}_tp_{xy}. \tag{22·1}$$

A joint-life annuity to (x) and (y) is an annuity payable so long as both of them survive. The value of such an annuity of 1, payable yearly, is

$$a_{xy} = \frac{1}{l_{xy}} \sum_{t=1}^{\infty} v^t l_{x+t:y+t}. \tag{22·2}$$

The value of the corresponding annuity payable continuously is

$$\bar{a}_{xy} = \frac{1}{l_{xy}} \int_0^{\infty} v^t l_{x+t:y+t}\,dt. \tag{22·3}$$

Corresponding to the relationship between a_x and $\bar{a}_x$ we have

$$\bar{a}_{xy} = a_{xy} + \tfrac{1}{2} - \tfrac{1}{12}(\mu_{xy} + \delta). \tag{22·4}$$

For assurances we have

$$A_{xy} = \frac{1}{l_{xy}} \sum_{t=1}^{\infty} v^t d_{x+t-1:y+t-1} \tag{22·5}$$

and

$$\bar{A}_{xy} = \frac{1}{l_{xy}} \int_0^{\infty} v^t l_{x+t:y+t}\,\mu_{x+t:y+t}\,dt. \tag{22·6}$$

The expressions for temporary annuities, temporary assurances and endowment assurances are also similar to those in respect of single lives. The symbol for a joint-life temporary assurance payable at the end of the year of death is ${}_nA_{xy}$ or $A^{\;1}_{xy:\overline{n}|}$. Thus

$$a_{xy:\overline{n}|} = a_{xy} - {}_nE_{xy}\, a_{x+n:y+n},$$

$${}_nA_{xy} = \frac{1}{l_{xy}} \sum_{t=1}^{n} v^t d_{x+t-1:y+t-1}$$

and
$$A_{xy:\overline{n}|} = \frac{1}{l_{xy}} \left\{ \sum_{t=1}^{n} v^t d_{x+t-1:y+t-1} + v^n l_{x+n:y+n} \right\}.$$

The premium conversion relationships developed for single-life assurances apply equally to joint-life assurances. Thus

$$A_{xy:\overline{n}|} = 1 - d\ddot{a}_{xy:\overline{n}|}, \tag{22·7}$$

$$P_{xy:\overline{n}|} = \frac{1}{\ddot{a}_{xy:\overline{n}|}} - d, \tag{22·8}$$

$$\bar{A}_{xy:\overline{n}|} = 1 - \delta\bar{a}_{xy:\overline{n}|} \tag{22·9}$$

and
$$\bar{P}(\bar{A}_{xy:\overline{n}|}) = \frac{1}{\bar{a}_{xy:\overline{n}|}} - \delta. \tag{22·10}$$

The corresponding formulae without $\overline{n}|$ for whole-life assurances also apply.

It will be seen in Chapter 23 that the premium conversion formulae have an even wider application.

22·2. Last survivor probabilities, annuities and assurances

This book has so far been concerned with probabilities, annuities and assurances which can be expressed in terms of a hypothetical function, l_x, $(al)_x$ or l_{xy}, in respect of which the force of decrement at duration t is predetermined, being μ_{x+t} for the mortality table, $(a\mu)_{x+t}$ for the multiple-decrement table and $\mu_{x+t:y+t}$ for the joint-life mortality table. The methods which have been developed have depended on the existence of a formula of the type

$${}_np = \exp\left\{ -\int_0^n \mu_{+t}\, dt \right\}.$$

We have now to consider the probability that at least one of two lives, (x) and (y), will survive n years. This probability cannot be expressed in the same way, because the force of decrement at duration t is not predetermined, but depends on whether both lives, or (x) only, or (y) only, are alive at duration t. As will be seen in the next paragraph, however, the probability in question can easily be expressed in terms of single-life and joint-life probabilities on which values of annuities and assurances can be based.

The probability that at least one of two lives, (x) and (y), will survive n years is represented by the symbol ${}_np_{\overline{xy}}$. This probability is the sum of the three probabilities

(i) that (x) will survive, (y) having died,

(ii) that (y) will survive, (x) having died,

(iii) that both lives will survive.

Assuming that the probabilities of survival of the two lives are independent, we have

$$\begin{aligned} {}_np_{\overline{xy}} &= (1-{}_np_y)\,{}_np_x + (1-{}_np_x)\,{}_np_y + {}_np_{xy} \\ &= {}_np_x + {}_np_y - {}_np_{xy}. \end{aligned} \tag{22·11}$$

Alternatively the probability may be written down as the complement of the probability that neither of the lives will survive n years, i.e.

$$\begin{aligned} {}_np_{\overline{xy}} &= 1-(1-{}_np_x)(1-{}_np_y) \\ &= {}_np_x + {}_np_y - {}_np_{xy} \end{aligned}$$

as before.

Similarly an annuity of 1 payable yearly until the death of the survivor of two lives is represented by $a_{\overline{xy}}$. This type of annuity is known as a last survivor annuity.

$$\begin{aligned} a_{\overline{xy}} &= \sum_{t=1}^{\infty} v^t\,{}_tp_{\overline{xy}} \\ &= \sum_{t=1}^{\infty} v^t\,{}_tp_x + \sum_{t=1}^{\infty} v^t\,{}_tp_y - \sum_{t=1}^{\infty} v^t\,{}_tp_{xy}. \end{aligned}$$

$$\therefore \quad a_{\overline{xy}} = a_x + a_y - a_{xy}. \tag{22·12}$$

Formulae (22·11) and (22·12) can also be obtained directly by general reasoning.

An assurance payable on the last death of (x) and (y) is equivalent to an assurance payable on the death of (x) together with an assurance payable on the death of (y) less an assurance payable on the first death of (*x*) and (*y*). Such an assurance is represented by the symbol $\bar{A}_{\overline{xy}}$, and is known as a last survivor assurance.

$$\bar{A}_{\overline{xy}} = \bar{A}_x + \bar{A}_y - \bar{A}_{xy}.$$

Similarly, if the sum assured is payable at the end of the year of death,

$$A_{\overline{xy}} = A_x + A_y - A_{xy}. \tag{22·13}$$

Using the premium conversion relationships, we have

$$A_{\overline{xy}} = (1 - d\ddot{a}_x) + (1 - d\ddot{a}_y) - (1 - d\ddot{a}_{xy})$$

$$= 1 - d(\ddot{a}_x + \ddot{a}_y - \ddot{a}_{xy}).$$

$$\therefore \quad A_{\overline{xy}} = 1 - d\ddot{a}_{\overline{xy}}. \tag{22·14}$$

The annual premium for a last survivor assurance (the sum assured being payable at the end of the year of death) is

$$P_{\overline{xy}} = \frac{A_{\overline{xy}}}{\ddot{a}_{\overline{xy}}} = \frac{A_x + A_y - A_{xy}}{\ddot{a}_x + \ddot{a}_y - \ddot{a}_{xy}}.$$

Also

$$P_{\overline{xy}} = \frac{A_{\overline{xy}}}{\ddot{a}_{\overline{xy}}} = \frac{1 - d\ddot{a}_{\overline{xy}}}{\ddot{a}_{\overline{xy}}}.$$

$$\therefore \quad P_{\overline{xy}} = \frac{1}{\ddot{a}_{\overline{xy}}} - d. \tag{22·15}$$

It will be seen that it is not correct to write $P_{\overline{xy}} = P_x + P_y - P_{xy}$.

When there are three lives (*x*), (*y*) and (*z*), formula (22·11) becomes

$${}_n p_{\overline{xyz}} = 1 - (1 - {}_n p_x)(1 - {}_n p_y)(1 - {}_n p_z)$$

$$= {}_n p_x + {}_n p_y + {}_n p_z - {}_n p_{xy} - {}_n p_{xz} - {}_n p_{yz} + {}_n p_{xyz}. \tag{22·16}$$

Formulae (22·12) and (22·13) may be modified similarly.

Formulae (22·14) and (22·15) show that the premium conversion relationships apply, as might have been expected, to last survivor assurances.

The expressions for temporary annuities, temporary assurances and endowment assurances are:

$$a_{\overline{xy}:\overline{n}|} = a_{x:\overline{n}|} + a_{y:\overline{n}|} - a_{xy:\overline{n}|},$$

$${}_nA_{\overline{xy}} = A^1_{x:\overline{n}|} + A^1_{y:\overline{n}|} - {}_nA_{xy}$$

and

$$A_{\overline{xy}:\overline{n}|} = A_{x:\overline{n}|} + A_{y:\overline{n}|} - A_{xy:\overline{n}|}.$$

22·3. The probability of survival of exactly r lives

The symbol ${}_n p_{\overline{xyz\ldots(m)}}^{[r]}$ is used to represent the probability that out of m lives, (x), (y), (z), etc., exactly r lives will survive n years. On the assumption that the individual probabilities of survival are independent, this probability is the coefficient of a^r in the expansion

$$({}_n p_x a + {}_n q_x)({}_n p_y a + {}_n q_y)({}_n p_z a + {}_n q_z)\ldots.$$

Since ${}_n q_x = 1 - {}_n p_x$, etc., this expression can be written

$$\{1 + {}_n p_x(a-1)\}\{1 + {}_n p_y(a-1)\}\{1 + {}_n p_z(a-1)\}\ldots$$
$$= 1 + (\Sigma {}_n p_x)(a-1) + (\Sigma {}_n p_{xy})(a-1)^2 + \ldots + (\Sigma {}_n p_{xyz\ldots(m)})(a-1)^m.$$

If we write Z_1 for $\Sigma {}_n p_x$, Z_2 for $\Sigma {}_n p_{xy}$ and so on, this becomes

$$1 + Z_1(a-1) + Z_2(a-1)^2 + \ldots + Z_m(a-1)^m.$$

The coefficient of a^r in this expression is

$$Z_r - (r+1) Z_{r+1} + \binom{r+2}{2} Z_{r+2} - \binom{r+3}{3} Z_{r+3} + \ldots + (-1)^{m-r}\binom{m}{m-r} Z_m.$$

$$\therefore \quad {}_n p_{\overline{xyz\ldots(m)}}^{[r]} = Z_r - (r+1) Z_{r+1} + \binom{r+2}{2} Z_{r+2} - \binom{r+3}{3} Z_{r+3} + \ldots$$
$$+ (-1)^{m-r}\binom{m}{m-r} Z_m. \qquad (22\cdot17)$$

Now, by the binomial theorem,

$$\frac{Z^r}{(1+Z)^{r+1}} = Z^r - (r+1) Z^{r+1} + \binom{r+2}{2} Z^{r+2} - \binom{r+3}{3} Z^{r+3} + \ldots.$$

If, in the expression on the right-hand side of this relationship, we replace Z^r, Z^{r+1}, ... by Z_r, Z_{r+1}, ... and regard Z_{m+1}, Z_{m+2}, ... as being equal to zero, the expression becomes identical with the right-hand side of formula (22·17). It follows that formula (22·17) can be expressed symbolically in the concise form

$${}_n p_{\overline{xyz\ldots(m)}}^{[r]} = \frac{Z^r}{(1+Z)^{r+1}}, \qquad (22\cdot18)$$

it being understood that the expression on the right-hand side of this formula will be expanded by the binomial theorem as far as the term containing Z^m (subsequent terms being ignored) and that Z^r, Z^{r+1}, ... in the expansion will be replaced by Z_r, Z_{r+1}, ...

As an example of the application of formula (22·18), let us assume that an expression is required for the probability that of two lives, (x) and (y), exactly one life will survive n years.

Here $m=2$ and $r=1$.

$$Z_1={}_n p_x+{}_n p_y,\quad Z_2={}_n p_{xy}.$$

$$\frac{Z}{(1+Z)^2}=Z-2Z^2+\dots.$$

$$\therefore\quad {}_n p_{\overline{xy}}^{[1]}=Z_1-2Z_2$$

$$={}_n p_x+{}_n p_y-2{}_n p_{xy}.$$

As a further example let us assume that an expression is required for the probability that of four lives, (x), (y), (z) and (w), exactly two will survive n years.

Here $m=4$ and $r=2$.

$$Z_1={}_n p_x+{}_n p_y+{}_n p_z+{}_n p_w,$$

$$Z_2={}_n p_{xy}+{}_n p_{xz}+{}_n p_{xw}+{}_n p_{yz}+{}_n p_{yw}+{}_n p_{zw},$$

$$Z_3={}_n p_{xyz}+{}_n p_{xyw}+{}_n p_{xzw}+{}_n p_{yzw},$$

$$Z_4={}_n p_{xyzw}.$$

$$\frac{Z^2}{(1+Z)^3}=Z^2-3Z^3+6Z^4+\dots.$$

$$\therefore\quad {}_n p_{\overline{xyzw}}^{[2]}=Z_2-3Z_3+6Z_4$$

$$={}_n p_{xy}+{}_n p_{xz}+{}_n p_{xw}+{}_n p_{yz}+{}_n p_{yw}+{}_n p_{zw}$$

$$-3({}_n p_{xyz}+{}_n p_{xyw}+{}_n p_{xzw}+{}_n p_{yzw})+6{}_n p_{xyzw}.$$

22·4. The probability of survival of at least *r* lives

The probability that out of m lives, (x), (y), (z), etc., at least r lives will survive n years is represented by ${}_n p_{\overline{xyz\dots(m)}}^{r}$, i.e. with no brackets round the r. When $r=1$ it is omitted from the symbol.

$${}_n p_{\overline{xyz\dots(m)}}^{r}={}_n p_{\overline{xyz\dots(m)}}^{[r]}+{}_n p_{\overline{xyz\dots(m)}}^{[r+1]}+\dots. \qquad (22\cdot19)$$

Re-writing the right-hand side in terms of Z_r, Z_{r+1}, ... with the aid of formula (22·17), we have

$${}_n p_{\overline{xyz\dots(m)}}^{r}=Z_r-rZ_{r+1}+\binom{r+1}{2}Z_{r+2}-\binom{r+2}{3}Z_{r+3}+\dots$$
$$+(-1)^{m-r}\binom{m-1}{m-r}Z_m. \qquad (22\cdot20)$$

It will be seen that this formula can be expressed symbolically in the form

$${}_n p_{\overline{xyz\dots(m)}}^{r}=\frac{Z^r}{(1+Z)^r}, \qquad (22\cdot21)$$

it being understood, as in the case of formula (22·18), that the expression on the right-hand side will be expanded by the binomial theorem as far as the term containing Z^m and that Z^r, Z^{r+1}, ... in the expansion will be replaced by Z_r, Z_{r+1}, ...

The practical application of formula (22·21) may be illustrated by using it to obtain an expression for the probability that of four lives, (x), (y), (z) and (w), at least two will survive n years.

Here $m=4$ and $r=2$.

$$\frac{Z^2}{(1+Z)^2}=Z^2-2Z^3+3Z^4+\dots.$$

$$\therefore\quad {}_np_{\overline{xyzw}}^{\;2}=Z_2-2Z_3+3Z_4$$

$$={}_np_{xy}+{}_np_{xz}+{}_np_{xw}+{}_np_{yz}+{}_np_{yw}+{}_np_{zw}$$
$$-2({}_np_{xyz}+{}_np_{xyw}+{}_np_{xzw}+{}_np_{yzw})+3{}_np_{xyzw}.$$

Formulae (22·11) and (22·16) can easily be reproduced by this method, by putting $m=2$ and 3 respectively and $r=1$.

22·5. Annuities depending on the survival of at least r lives

The symbol $a_{\overline{xyz\dots(m)}}^{\;r}$ is used to represent an annuity payable so long as at least r lives survive.

Since

$$a_{\overline{xyz\dots(m)}}^{\;r}=\sum_{t=1}^{\infty}v^t\,{}_tp_{\overline{xyz\dots(m)}}^{\;r},$$

the method of the previous section can be immediately extended to such an annuity. Thus

$$a_{\overline{xyz\dots(m)}}^{\;r}=Z_r-rZ_{r+1}+\binom{r+1}{2}Z_{r+2}-\binom{r+2}{3}Z_{r+3}+\dots$$
$$+(-1)^{m-r}\binom{m-1}{m-r}Z_m,\qquad(22\cdot22)$$

where Z_t denotes $\Sigma a_{xyz\dots(t)}$. This can be expressed symbolically in the form

$$a_{\overline{xyz\dots(m)}}^{\;r}=\frac{Z^r}{(1+Z)^r},\qquad(22\cdot23)$$

the right-hand side being interpreted as before.

As an illustration of the use of formula (22·23), let us consider an annuity of 1 payable yearly so long as at least three lives survive out of four lives, (x), (y), (z) and (w).

Here $m=4$ and $r=3$.

$$\frac{Z^3}{(1+Z)^3}=Z^3-3Z^4+\dots.$$

$$\therefore \quad a_{\overline{xyzw}}^{\;3}=Z_3-3Z_4$$

$$=a_{xyz}+a_{xyw}+a_{xzw}+a_{yzw}-3a_{xyzw}.$$

By putting $m=2$ and $r=1$ in formula (22·22) or (22·23), we have formula (22·12).

22·6. Selection

If select mortality is assumed, the formulae developed above will apply with obvious modifications which need hardly be demonstrated here.

22·7. Office premiums

Office premiums for joint-life and last survivor assurances are usually calculated on the basis of select mortality by a formula similar to that used for single-life assurances. The loadings included in the formula will not necessarily be the same as those which are used for single-life assurances, since the office may wish to increase the allowance for expenses and the margin for contingencies (see section 21·6).

22·8. Policy values

Policy values are developed for joint-life assurances in exactly the same manner as for single-life assurances. Thus

$${}_tV_{xy}=A_{x+t:y+t}-P_{xy}\ddot{a}_{x+t:y+t}$$

$$=1-\frac{\ddot{a}_{x+t:y+t}}{\ddot{a}_{xy}}.$$

For a last survivor assurance an additional complication arises as either one or both lives may survive to the duration at which the value is required.

Thus if both lives survive

$${}_tV_{\overline{xy}}=A_{\overline{x+t:y+t}}-P_{\overline{xy}}\ddot{a}_{\overline{x+t:y+t}}$$

$$=1-\frac{\ddot{a}_{\overline{x+t:y+t}}}{\ddot{a}_{\overline{xy}}},$$

but if only one life, (x), survives the value becomes

$$A_{x+t}-P_{\overline{xy}}\ddot{a}_{x+t}=1-\frac{\ddot{a}_{x+t}}{\ddot{a}_{\overline{xy}}}.$$

22·9. Evaluation of joint-life functions

A discussion of the evaluation of joint-life functions will be deferred to Chapters 26 and 27.

EXAMPLES

Example 22·1

If $\mu_x=\dfrac{1}{100-x}$ calculate the complete last survivor expectation of life of two lives aged 40 and 50.

Solution

$$
\begin{aligned}
{}_np_x &= \exp\left\{-\int_0^n \mu_{x+t}\,dt\right\} \\
&= \exp\{\log_e(100-x-n)-\log_e(100-x)\} \\
&= \frac{100-(x+n)}{100-x} \\
&= 1-\frac{n}{100-x}
\end{aligned}
$$

$$\mathring{e}_x=\int_0^{100-x}\left(1-\frac{n}{100-x}\right)dn=\frac{100-x}{2}.$$

$$
\begin{aligned}
\mathring{e}_{xy}\,(x>y) &= \int_0^{100-x}\left(1-\frac{n}{100-x}\right)\left(1-\frac{n}{100-y}\right)dn \\
&= \frac{100-x}{2}-\frac{(100-x)^2}{6(100-y)}.
\end{aligned}
$$

$$\therefore\quad \mathring{e}_{50}=25,\quad \mathring{e}_{40}=30\quad\text{and}\quad \mathring{e}_{40:50}=25-\frac{2500}{360}=18{\cdot}06.$$

$$\therefore\quad \mathring{e}_{\overline{40:50}}=25+30-18{\cdot}06=36{\cdot}94.$$

Example 22·2

Find an expression, as far as possible in terms of whole-life deferred annuities (single and joint), for the single premium for a deferred annuity in the case of three lives now aged 40, 45 and 50; the annuity to be payable so long as at least one of them is alive and aged over 60. The annuity is to be £300 per annum while exactly one is over 60, £400 per annum while exactly two are over 60, and £500 per annum while all three are over 60.

Solution

Let us use the Z method and divide by £100. We have to provide

3 while at least 1 is over 60	$\dfrac{3Z}{1+Z}=3Z-3Z^2+3Z^3-\dots$
plus 1 while at least 2 are over 60	$\dfrac{Z^2}{(1+Z)^2}= \quad Z^2-2Z^3+\dots$
plus 1 while at least 3 are over 60	$\dfrac{Z^3}{(1+Z)^3}= \quad Z^3-\dots$
	Total $3Z-2Z^2+2Z^3-\dots$

The answer is therefore

$$
\begin{aligned}
&£300({}_{10|}a_{50}+{}_{15|}a_{45}+{}_{20|}a_{40})\\
&-£200({}_{15|}a_{45:50}+{}_{20|}a_{40:45}+{}_{20|}a_{40:50})\\
&+£200\ {}_{20|}a_{40:50:60}.
\end{aligned}
$$

Example 22·3

Find by the Z method expressions for the probabilities that of three lives aged respectively 30, 35 and 40,

(*a*) One and one only will die between ages 45 and 50,

(*b*) At least one will die between ages 45 and 50.

Test your results by an independent method and show how to evaluate the expressions with a minimum of arithmetical labour.

Solution

Write x for $\dfrac{l_{45}-l_{50}}{l_{30}}$, y for $\dfrac{l_{45}-l_{50}}{l_{35}}$ and z for $\dfrac{l_{45}-l_{50}}{l_{40}}$. Then

$$
\begin{aligned}
(a)&=Z_1-2Z_2+3Z_3\\
&=(x+y+z)-2(xy+xz+yz)+3xyz \qquad (1)
\end{aligned}
$$

and $$(b) = Z_1 - Z_2 + Z_3$$
$$= (x+y+z) - (xy+xz+yz) + xyz. \quad (2)$$

Alternatively

$$(a) = x(1-y)(1-z) + y(1-x)(1-z) + z(1-x)(1-y)$$

and $$(b) = 1-(1-x)(1-y)(1-z).$$

These expressions can be reduced to (1) and (2) respectively.

The original expressions (1) and (2) are in the most convenient form for calculation. The best way to proceed is as follows:

First calculate x, y and z.

Then obtain xy, xz and yz by multiplication.

Then obtain xyz.

Then calculate $(x+y+z)$ and $(xy+xz+yz)$ from which (1) and (2) can easily be obtained.

This method involves the minimum of labour, since, apart from the multiplication of $(xy+xz+yz)$ by 2 and xyz by 3, which need hardly be considered, only three divisions and four multiplications are required.

CHAPTER 23

SOME GENERAL THEOREMS OF LIFE CONTINGENCIES

23·1. Introduction

General theorems of life contingencies are obtained by the development of general expressions for probabilities and for the values of benefits which depend on the existence or termination of a given status.

In the formulae which have so far been developed in this book, the status has been specified, or at least implied. For example, the annuity a_x, considered in Volume I, is payable during the continuance of the simple status '(x) being alive'. The annuity $a_{x:\overline{n}|}$, also considered in Volume I, is payable during the status '(x) being alive and a term of n years not having expired'. In Volume II we have to consider joint-life statuses (such as '(x) and (y) both being alive'), last survivor statuses (such as 'at least one of the two lives (x) and (y) being alive') and other more complicated statuses.

In the present chapter, it is proposed to develop some general formulae which apply to a status the precise terms of which are not specified. These formulae have a twofold value. In the first place, as will be seen in the course of the chapter, they are instructive since they show that formulae which have already been developed in respect of particular statuses are special cases of formulae which are of wider application. In the second place, some of the general formulae can be used (and will in fact be used in the later sections of this chapter) for deriving useful approximate methods for dealing with various practical problems.

Only a superficial treatment of the subject will be attempted and consideration will be confined to statuses which exist continuously from the present time until they come to an end. Thus a status such as '(y) being alive and (x) having died' will not be included here. Most of the demonstrations will be based on continuous functions; the corresponding yearly proofs will not be given but the final expressions in terms of yearly functions will sometimes be shown. For convenience, selection will be ignored throughout the chapter, but the methods and formulae which will be developed are equally applicable to select tables.

23·2. Notation

In the development of general expressions in this chapter, a subscript placed at the lower right-hand corner of a main symbol indicates time. The origin from which time is measured will usually be obvious from the context. Thus l_t represents the number of groups* of lives whose status is expected to remain in force until time t out of l_0 groups of lives (all similar to one another) which are in existence at time 0; δ_t is the force of interest at time t, $\bar{P}_t$ is the premium per annum payable at duration t in respect of a policy which is subject to a variable continuous premium; and so on. This system of notation is not quite in accordance with the International Actuarial Notation, but it has the merit of being very convenient for the present purpose. Sometime it is desirable to place the subscript to the left of the main symbol, instead of the right, in order to avoid misunderstanding. For example, the probability that the status of a group, taken at random at time 0, will still be in force at time t is represented by the symbol ${}_tp$; the symbol p_t must be taken to mean the probability that the status of a group which is in existence at time t will remain in force until time $t+1$. Strictly speaking, the policy value at duration t of a policy subject to a continuous premium should be written ${}_t\bar{V}$ rather than $\bar{V}_t$; however, as will be seen from the development in a later section of this chapter, the use of the symbol $\bar{V}_t$ sometimes makes for neatness and does not cause confusion.

23·3. Probabilities and forces

It follows from the definition of l_t given in the previous section that the probability that the status of a group, taken at random at time 0, will still be in force at time t is

$$ {}_tp = l_t/l_0. \qquad (23\cdot1)$$

* The word 'group' is used here to mean one life, or two or more lives, according to the status. For example, in the case of a two-life last survivor status, the 'group' consists of two lives (x) and (y); the status will still be in force at time t if either (x) or (y) is then alive.

The force of decrement by mode α is μ_t^α and the force of decrement by all modes combined is μ_t.

$$\mu_t = \Sigma \mu_t^\alpha \tag{23·2}$$

and

$$\mu_t = -\frac{1}{l_t}\frac{dl_t}{dt}. \tag{23·3}$$

Some care must be used in interpreting the exact meaning of μ in more complicated cases. For example if a two-life last survivor status is involved

$$l_t/l_0 = {}_tp_x + {}_tp_y - {}_tp_{xy},$$

and μ_t would be the force of decrement for a large number of initially similar groups; μ_t, however, cannot be readily expressed in terms of μ_{x+t} and μ_{y+t}.

Formula (23·3) may be written

$$\mu_t = -\frac{d}{dt}\log_e l_t,$$

whence from formula (23·1)

$$_np = \exp\left\{-\int_0^n \mu_t dt\right\}. \tag{23·4}$$

23·4. Annuities

If δ_t is the force of interest at time t, the present value of 1 due n years hence is

$$\exp\left\{-\int_0^n \delta_t dt\right\}.$$

Hence the value of a pure endowment of 1, payable at the end of n years if the status is still in force, is

$$_nE = \exp\left\{-\int_0^n (\mu_t + \delta_t)\, dt\right\}. \tag{23·5}$$

The value of a continuous variable annuity, payable during the continuance of the status, under which the payment at time t is at the rate of $\phi(t)$ per annum, is

$$(v\bar{a}) = \int_0^\infty \phi(n) \exp\left\{-\int_0^n (\mu_t + \delta_t)\, dt\right\} dn. \tag{23·6}$$

The corresponding formula for a variable annuity payable yearly would be

$$(va) = \sum_{n=1}^{\infty} \phi(n) \prod_{t=0}^{n-1} \{v_t(1 - q_t)\}.$$

In the expression on the right-hand side of this formula, $\phi(n)$ is the payment at time n, v_t is equal to $1/(1+i_t)$, where i_t is the rate of interest during the $(t+1)$th year, q_t is the total rate of decrement during the $(t+1)$th year and

$$\prod_{t=0}^{n-1} f(t) = f(0) \times f(1) \times \ldots \times f(n-1).$$

In Chapter 12 it was shown that for a single-life annuity a constant addition to the force of mortality is equivalent to a constant addition to the force of interest. This is immediately apparent from formula (23·6), from which we see that any addition (constant throughout or varying with the duration) to the total force of decrement is equivalent to the same addition to the force of interest.

For example, if it is desired to calculate the value of a 20-year increasing joint-life annuity, where one life is subject to extra mortality equivalent to an addition of ·005 to μ for the first ten years and the other life is subject to extra mortality equivalent to an addition of ·005 to μ throughout the term, the result can be obtained by using normal mortality and assuming an addition of ·01 to the force of interest for the first ten years and an addition of ·005 thereafter.

23·5. Assurances

If the sum assured payable immediately on termination of the status at duration t by the mode of decrement α is $\bar{S}_t^\alpha$, by mode β $\bar{S}_t^\beta$, etc., then the value of an assurance for these sums is

$$(v\bar{A}) = \int_0^\infty \exp\left\{-\int_0^n (\mu_t + \delta_t)\, dt\right\} \Sigma(\bar{S}_n^\alpha \mu_n^\alpha)\, dn, \qquad (23·7)$$

where the summation is over all modes of decrement.

If the assurance is subject to a variable continuous premium $\bar{P}_t$, then the value of an annuity of this premium, obtained by writing $\bar{P}_n$ for $\phi(n)$ in formula (23·6), must be equal to $(v\bar{A})$.

If the assurance is subject to a level continuous premium $\bar{P}$ then

$$\bar{P} = \frac{\int_0^\infty \exp\left\{-\int_0^n (\mu_t + \delta_t)\, dt\right\} \Sigma(\bar{S}_n^\alpha \mu_n^\alpha)\, dn}{\int_0^\infty \exp\left\{-\int_0^n (\mu_t + \delta_t)\, dt\right\} dn}. \qquad (23·8)$$

Let $\bar{P}^{\alpha\beta}(\bar{S}_t^\alpha . \bar{S}_t^\beta)$ denote the level continuous premium, payable throughout the continuance of a status which is subject to two modes of decrement, to provide sums assured of $\bar{S}_t^\alpha$ and $\bar{S}_t^\beta$. Then, from formula (23·8) or from general reasoning,

$$\bar{P}^{\alpha\beta}(\bar{S}_t^\alpha . \bar{S}_t^\beta) = \bar{P}^{\alpha\beta}(\bar{S}_t^\alpha . 0) + \bar{P}^{\alpha\beta}(0 . \bar{S}_t^\beta). \tag{23·9}$$

Similarly if $\bar{P}^{\alpha\beta\gamma}(\bar{S}_t^\alpha . \bar{S}_t^\beta . \bar{S}_t^\gamma)$ denotes the premium, payable throughout the continuance of a status which is subject to three modes of decrement, to provide sums assured of $\bar{S}_t^\alpha$, $\bar{S}_t^\beta$ and $\bar{S}_t^\gamma$,

$$\bar{P}^{\alpha\beta\gamma}(\bar{S}_t^\alpha . \bar{S}_t^\beta . \bar{S}_t^\gamma) = \bar{P}^{\alpha\beta\gamma}(\bar{S}_t^\alpha . 0 . 0) + \bar{P}^{\alpha\beta\gamma}(0 . \bar{S}_t^\beta . 0) + \bar{P}^{\alpha\beta\gamma}(0 . 0 . \bar{S}_t^\gamma). \tag{23·10}$$

Formulae (23·9) and (23·10) express, in general terms, various relationships which have already been developed, directly or indirectly, in this book. For example if we put

$$\bar{S}_t^\alpha = 1,$$
$$\bar{S}_t^\beta = 1,$$
$$\mu_t^\alpha = \mu_{x+t}$$

and

$$\mu_t^\beta = 0, \quad t < n,$$
$$= \infty, \quad t = n,$$

formula (23·9) becomes

$$\bar{P}(\bar{A}_{x:\overline{n}|}) = \bar{P}(\bar{A}^{\,1}_{x:\overline{n}|}) + \bar{P}(A_{x:\overset{\,1}{\overline{n}|}}).$$

Again, if we put

$$\bar{S}_t^\alpha = 1,$$
$$\bar{S}_t^\beta = 1,$$
$$\bar{S}_t^\gamma = 1,$$
$$\mu_t^\alpha = \mu_{x+t},$$
$$\mu_t^\beta = \mu_{y+t}$$

and

$$\mu_t^\gamma = \mu_{z+t},$$

formula (23·10) becomes

$$\bar{P}(\bar{A}_{xyz}) = \bar{P}(\bar{A}^{\,1}_{xyz}) + \bar{P}(\bar{A}_{x\overset{1}{y}z}) + \bar{P}(\bar{A}_{xy\overset{1}{z}}).$$

Formulae (23·9) and (23·10) are themselves special cases of a more general formula

$$\bar{P}^{\alpha\beta\cdots}(\bar{S}_t^\alpha . \bar{S}_t^\beta \ldots) = \bar{P}^{\alpha\beta\cdots}({}^1\bar{S}_t^\alpha . {}^1\bar{S}_t^\beta \ldots) + \bar{P}^{\alpha\beta\cdots}({}^2\bar{S}_t^\alpha . {}^2\bar{S}_t^\beta \ldots) + \bar{P}^{\alpha\beta\cdots}({}^3\bar{S}_t^\alpha . {}^3\bar{S}_t^\beta \ldots) + \ldots, \tag{23·11}$$

where $$\bar{S}_t^\alpha = {}^1\bar{S}_t^\alpha + {}^2\bar{S}_t^\alpha + {}^3\bar{S}_t^\alpha + \ldots,$$

$$\bar{S}_t^\beta = {}^1\bar{S}_t^\beta + {}^2\bar{S}_t^\beta + {}^3\bar{S}_t^\beta + \ldots$$

and so on.

We shall have occasion to use formula (23·11) later in this chapter.

23·6. Premium conversion relationships

It is assumed in this section that

(1) the sum assured is constant and independent of the mode of decrement,

(2) the rate of interest is constant, and

(3) annual premiums, if payable, are at a level rate.

Then from formula (23·6)

$$\bar{a} = \int_0^\infty \exp\left\{-\int_0^n (\mu_t + \delta)\,dt\right\} dn$$

and from formula (23·7)

$$\bar{A} = \int_0^\infty \exp\left\{-\int_0^n (\mu_t + \delta)\,dt\right\} \mu_n\,dn.$$

Now

$$\frac{d}{dn}\exp\left\{-\int_0^n (\mu_t + \delta)\,dt\right\} = -\exp\left\{-\int_0^n (\mu_t + \delta)\,dt\right\}(\mu_n + \delta).$$

Integrating between the limits 0 and ∞ we have

$$\left[\exp\left\{-\int_0^n (\mu_t + \delta)\,dt\right\}\right]_0^\infty = -\bar{A} - \delta\bar{a}.$$

$$\therefore\quad \exp\left\{-\int_0^\infty (\mu_t + \delta)\,dt\right\} - \exp\left\{-\int_0^0 (\mu_t + \delta)\,dt\right\} = -\bar{A} - \delta\bar{a},$$

i.e. $$0 - 1 = -\bar{A} - \delta\bar{a}.$$

$$\therefore\quad \bar{A} = 1 - \delta\bar{a}. \qquad (23·12)$$

Dividing by $\bar{a}$, we have

$$\bar{P}(\bar{A}) = \frac{1}{\bar{a}} - \delta. \qquad (23·13)$$

It will be noted that these expressions hold whenever the conditions specified at the beginning of this section are fulfilled. For example they apply to a single-life, joint-life or last survivor status. Again by putting

$$\mu_t^\alpha = \mu_{x+t}$$

and

$$\mu_t^\beta = 0, \quad t < n,$$
$$= \infty, \quad t = n,$$

we see that they apply also to an endowment assurance.

If expressions are developed on a yearly basis, the corresponding formulae

$$A = 1 - d\ddot{a}$$

and

$$P = \frac{1}{\ddot{a}} - d$$

result.

Formulae (23·12) and (23·13), and the corresponding yearly formulae, can easily be established by general reasoning as in section 5 of Chapter 6 (Volume I of this book).

The conditions under which the premium conversion relationships hold may be expressed as follows:

(1) the status is in force continuously from time 0 until it comes to an end,

(2) $\bar{a}$ is the value of a continuous annuity (or $\ddot{a}$ is the value of an annuity-due) of 1 payable throughout the continuance of the status, and

(3) $\bar{A}$ is the value of an assurance of 1 payable immediately on termination of the status (or A is the value of an assurance of 1 payable at the end of the year of termination of the status).

*23·7. Policy values

The value of a policy is the difference between the value of the future benefits and the value of the future premiums at the date of valuation. We shall limit our consideration to those statuses in which the policy value is a function of the duration only. This will exclude a last survivor status where the policy value is a function not only of the duration but also of the number and ages of the lives surviving.

The relationship between the policy values at successive intervals can be obtained by treating the policy value at duration t as a single premium to provide, with the premiums payable in the interval t to $t+\delta t$, the benefits payable in the interval and the policy value at the end of the interval. Let the policy value at duration t be $\overline{V}_t$. Then, as $\delta t \to 0$,

$$\overline{V}_t + \overline{P}_t \delta t \to \Sigma(\bar{S}_t^\alpha \mu_t^\alpha)\, \delta t + (1 - \Sigma \mu_t^\alpha\, \delta t - \delta_t \delta t)\, \overline{V}_{t+\delta t}.$$

$$\therefore \quad \overline{V}_{t+\delta t} - \overline{V}_t \to \overline{P}_t \delta t + \delta_t \delta t\, \overline{V}_{t+\delta t} - \Sigma\{(\bar{S}_t^\alpha - \overline{V}_{t+\delta t})\, \mu_t^\alpha\}\, \delta t.$$

Proceeding to the limit, we have

$$\frac{d\bar{V}_t}{dt}=\bar{P}_t+\delta_t\bar{V}_t-\Sigma\{(\bar{S}_t^\alpha-\bar{V}_t)\,\mu_t^\alpha\}. \qquad (23\cdot14)$$

The corresponding formula for yearly functions is

$$V_{t+1}-V_t=P_t+(V_t+P_t)\,i_t-\Sigma\{(S_t^\alpha-V_{t+1})\,q_t^\alpha\}. \qquad (23\cdot15)$$

Consider two tables, one subject to three modes of decrement, α, β and γ, and the other subject to two modes of decrement, β and γ. Let μ_t^β and μ_t^γ have the same values in the two tables and let all three forces of decrement be finite except at the point where $t=\omega$, at which point either μ_t^β or μ_t^γ, or both, will necessarily be infinite.* Let the reserves under the two tables be $\bar{V}_t^{\alpha\beta\gamma}$ and $\bar{V}_t^{\beta\gamma}$ respectively and let the sums assured for an assurance under the first table be $\bar{V}_t^{\alpha\beta\gamma}$, $\bar{S}_t^\beta$ and $\bar{S}_t^\gamma$ and for an assurance under the second table $\bar{S}_t^\beta$ and $\bar{S}_t^\gamma$. Then the premiums for assurances under the two tables are respectively

$$\bar{P}^{\alpha\beta\gamma}(\bar{V}_t^{\alpha\beta\gamma}.\bar{S}_t^\beta.\bar{S}_t^\gamma)=\bar{P}^{\alpha\beta\gamma}, \quad \text{say,}$$

and

$$\bar{P}^{\beta\gamma}(\bar{S}_t^\beta.\bar{S}_t^\gamma)=\bar{P}^{\beta\gamma}, \quad \text{say.}$$

From formula (23·14) it follows that

$$\frac{d\bar{V}_t^{\alpha\beta\gamma}}{dt}=\bar{P}^{\alpha\beta\gamma}+\delta_t\bar{V}_t^{\alpha\beta\gamma}-(\bar{S}_t^\beta-\bar{V}_t^{\alpha\beta\gamma})\,\mu_t^\beta-(\bar{S}_t^\gamma-\bar{V}_t^{\alpha\beta\gamma})\,\mu_t^\gamma$$

and

$$\frac{d\bar{V}_t^{\beta\gamma}}{dt}=\bar{P}^{\beta\gamma}+\delta_t\bar{V}_t^{\beta\gamma}-(\bar{S}_t^\beta-\bar{V}_t^{\beta\gamma})\,\mu_t^\beta-(\bar{S}_t^\gamma-\bar{V}_t^{\beta\gamma})\,\mu_t^\gamma.$$

These two differential equations in $\bar{V}_t^{\alpha\beta\gamma}$ and $\bar{V}_t^{\beta\gamma}$ are identical except for the constants $\bar{P}^{\alpha\beta\gamma}$ and $\bar{P}^{\beta\gamma}$. Now $\bar{V}_0^{\alpha\beta\gamma}=\bar{V}_0^{\beta\gamma}=0$ and $\bar{V}_\omega^{\alpha\beta\gamma}=\bar{V}_\omega^{\beta\gamma}=1$. It follows that in this special case $\bar{V}_t^{\alpha\beta\gamma}=\bar{V}_t^{\beta\gamma}$ for all values of t and $\bar{P}^{\alpha\beta\gamma}=\bar{P}^{\beta\gamma}$, or in full notation

$$\bar{P}^{\alpha\beta\gamma}(\bar{V}_t^{\beta\gamma}.\bar{S}_t^\beta.\bar{S}_t^\gamma)=\bar{P}^{\beta\gamma}(\bar{S}_t^\beta.\bar{S}_t^\gamma). \qquad (23\cdot16)$$

Similarly it can be shown that

$$\bar{P}^{\alpha\beta\gamma}(\bar{V}_t^\gamma.\bar{V}_t^\gamma.\bar{S}_t^\gamma)=\bar{P}^\gamma(\bar{S}_t^\gamma). \qquad (23\cdot17)$$

When there are only two modes of decrement, either of these formulae reduces to

$$\bar{P}^{\alpha\beta}(\bar{V}_t^\beta.\bar{S}_t^\beta)=\bar{P}^\beta(\bar{S}_t^\beta). \qquad (23\cdot18)$$

* It is assumed that the tables have a limiting point ω. If this is not the case the suffix ω, in the demonstration which follows, should be replaced by ∞.

This formula provides the mathematical justification for calculating the premiums for life assurance on a single-decrement mortality table despite the fact that policies may terminate by surrender as well as by death. The reasons why, in practice, surrender values are somewhat less than the reserves are explained briefly in section 12 of Chapter 11 (Volume I).

A simple illustration of formula (23·18) is provided by putting

$$\bar{S}_t^\beta = 1,$$

$$\mu_t^\alpha = \mu_{x+t}$$

and

$$\mu_t^\beta = 0, \quad t < n,$$

$$= \infty, \quad t = n.$$

The left-hand side of the formula then becomes the annual premium, payable continuously, for an n-year endowment assurance on (x) under which the sum assured is 1 on survival to the end of n years and $\bar{V}_{\overline{n}|}$ on previous death; the formula shows that this premium is equal to $\bar{P}_{\overline{n}|}$. This can be verified by general reasoning, because if a capital redemption policy were issued subject to the special condition that in the event of the death of (x) during the term the policy would be cancelled and the policy value would be paid, this condition (provided that all questions of loading were ignored) would make no difference to the premium.

23·8. Lidstone's approximation for joint-life endowment assurance premiums

As an interesting example of the application of formulae (23·11), (23·16) and (23·17) to a practical problem, we may consider the following approximate formula for the annual premium for a joint-life endowment assurance.

$$P_{xy:\overline{n}|} \doteq P_{x:\overline{n}|} + P_{y:\overline{n}|} - P_{\overline{n}|}. \tag{23·19}$$

This formula was first suggested by G. J. Lidstone in *J.I.A.* **33** [1897], 354.

Let

$$\mu_t^\alpha = \mu_{x+t},$$

$$\mu_t^\beta = \mu_{y+t}$$

and

$$\mu_t^\gamma = 0, \quad t < n,$$

$$= \infty, \quad t = n.$$

Then
$$\bar{P}^{\alpha\beta\gamma}(1.1.1)=\bar{P}(\bar{A}_{xy:\overline{n}|}),$$
$$\bar{P}^{\alpha\gamma}(1.1)=\bar{P}(\bar{A}_{x:\overline{n}|}),$$
$$\bar{P}^{\beta\gamma}(1.1)=\bar{P}(\bar{A}_{y:\overline{n}|})$$
and
$$\bar{P}^{\gamma}(1)=\bar{P}_{\overline{n}|}.$$

From formula (23·11)

$$\bar{P}^{\alpha\beta\gamma}(1.1.1)=\bar{P}^{\alpha\beta\gamma}(1.\bar{V}_t^{\alpha\gamma}.1)+\bar{P}^{\alpha\beta\gamma}(\bar{V}_t^{\beta\gamma}.1.1)$$
$$-\bar{P}^{\alpha\beta\gamma}(\bar{V}_t^{\gamma}.\bar{V}_t^{\gamma}.1)-\bar{P}^{\alpha\beta\gamma}\{(\bar{V}_t^{\beta\gamma}-\bar{V}_t^{\gamma}).(\bar{V}_t^{\alpha\gamma}-\bar{V}_t^{\gamma}).0\}.$$

But from formulae (23·16) and (23·17) the right-hand side of this formula is equal to

$$\bar{P}^{\alpha\gamma}(1.1)+\bar{P}^{\beta\gamma}(1.1)-\bar{P}^{\gamma}(1)-\bar{P}^{\alpha\beta\gamma}\{(\bar{V}_t^{\beta\gamma}-\bar{V}_t^{\gamma}).(\bar{V}_t^{\alpha\gamma}-\bar{V}_t^{\gamma}).0\}.$$

Returning to normal notation, we have

$$\bar{P}(\bar{A}_{xy:\overline{n}|})=\bar{P}(\bar{A}_{x:\overline{n}|})+\bar{P}(\bar{A}_{y:\overline{n}|})-\bar{P}_{\overline{n}|}-\bar{P}'^{\,1}_{\overline{xy}:\overline{n}|},$$

where $\bar{P}'^{\,1}_{\overline{xy}:\overline{n}|}$ is the premium for a joint-life temporary assurance under which the sum payable is

$$_t\bar{V}(\bar{A}_{y:\overline{n}|})-{}_t\bar{V}_{\overline{n}|}$$

if (x) dies first or
$$_t\bar{V}(\bar{A}_{x:\overline{n}|})-{}_t\bar{V}_{\overline{n}|}$$

if (y) dies first. This premium is very small in relation to $\bar{P}(\bar{A}_{xy:\overline{n}|})$. It follows therefore that

$$\bar{P}(\bar{A}_{xy:\overline{n}|})\fallingdotseq\bar{P}(\bar{A}_{x:\overline{n}|})+\bar{P}(\bar{A}_{y:\overline{n}|})-\bar{P}_{\overline{n}|}. \qquad (23{\cdot}20)$$

By altering the continuous functions to yearly ones, we have formula (23·19).

It is easy to verify formulae (23·19) and (23·20) by general reasoning. It is also easy to show that formula (23·19), in addition to containing an error corresponding to that in formula (23·20), overstates the values of $P_{xy:\overline{n}|}$ by a very small quantity equal to the premium for a temporary assurance payable only if both lives die in the same year during the term.

23·9. The extension of Lidstone's approximation to three or more lives

Formula (23·19) may be written

$$P_{xy:\overline{n}|}\fallingdotseq P_{\overline{n}|}+(P_{x:\overline{n}|}-P_{\overline{n}|})+(P_{y:\overline{n}|}-P_{\overline{n}|}). \qquad (23{\cdot}21)$$

Similarly when there are three lives we may write

$$P_{xyz:\overline{n|}} \doteqdot P_{\overline{n|}} + (P_{x:\overline{n|}} - P_{\overline{n|}}) + (P_{y:\overline{n|}} - P_{\overline{n|}}) + (P_{z:\overline{n|}} - P_{\overline{n|}}), \quad (23\cdot22)$$

and when there are four lives

$$P_{xyzw:\overline{n|}} \doteqdot P_{\overline{n|}} + (P_{x:\overline{n|}} - P_{\overline{n|}}) + (P_{y:\overline{n|}} - P_{\overline{n|}}) + (P_{z:\overline{n|}} - P_{\overline{n|}}) + (P_{w:\overline{n|}} - P_{\overline{n|}}). \quad (23\cdot23)$$

When there are three or more lives other variations are possible, for example

$$P_{xyz:\overline{n|}} \doteqdot P_{xy:\overline{n|}} + (P_{z:\overline{n|}} - P_{\overline{n|}}), \quad (23\cdot24)$$

$$P_{xyz:\overline{n|}} \doteqdot P_{xz:\overline{n|}} + P_{yz:\overline{n|}} - P_{z:\overline{n|}} \quad (23\cdot25)$$

and $$P_{xyz:\overline{n|}} \doteqdot P_{\overline{n|}} + (P_{xy:\overline{n|}} - P_{y:\overline{n|}}) + (P_{yz:\overline{n|}} - P_{z:\overline{n|}}) + (P_{xz:\overline{n|}} - P_{x:\overline{n|}}). \quad (23\cdot26)$$

23·10. The calculation of endowment assurance premiums when the life is subject to a certain type of extra mortality

Consider a life (x) subject to special mortality such that for all values of t between 0 and n

$$\mu'_{x+t} = (1+k)\,\mu_{x+t}. \quad (23\cdot27)$$

(The assumption that the special mortality takes this form is almost identical with assumption (c) in Volume I, section 12·3.)

Then, when $k=0$ $\quad P'_{x:\overline{n|}} = P_{x:\overline{n|}}$.

When $k=1$ $\quad P'_{x:\overline{n|}} = P_{xx:\overline{n|}}$

or $\quad P'_{x:\overline{n|}} = P_{x:\overline{n|}} + (P_{xx:\overline{n|}} - P_{x:\overline{n|}})$.

When $k=2$ $\quad P'_{x:\overline{n|}} = P_{xxx:\overline{n|}}$,

whence from formula (23·25)

$$P'_{x:\overline{n|}} \doteqdot 2P_{xx:\overline{n|}} - P_{x:\overline{n|}}$$

or $$P'_{x:\overline{n|}} \doteqdot P_{x:\overline{n|}} + 2(P_{xx:\overline{n|}} - P_{x:\overline{n|}}).$$

Hence we may say generally that

$$P'_{x:\overline{n|}} \doteqdot P_{x:\overline{n|}} + k(P_{xx:\overline{n|}} - P_{x:\overline{n|}}). \quad (23\cdot28)$$

From formula (23·19)

$$P_{xx:\overline{n|}} - P_{x:\overline{n|}} \doteqdot P_{x:\overline{n|}} - P_{\overline{n|}}.$$

Making this substitution in formula (23·28) we have

$$P'_{x:\overline{n|}} \doteqdot P_{x:\overline{n|}} + k(P_{x:\overline{n|}} - P_{\overline{n|}}). \quad (23\cdot29)$$

This formula provides a useful method of calculating approximately the net annual premium for an endowment assurance when the life is subject to special mortality of the type represented in formula (23·27).

Similarly, for a joint-life endowment assurance on two lives (x) and (y) who are subject to special mortality such that

$$\mu'_{x+t} = (1+k_1)\,\mu_{x+t}$$

and

$$\mu'_{y+t} = (1+k_2)\,\mu_{y+t},$$

we have

$$P'_{xy:\overline{n}|} \doteqdot P_{xy:\overline{n}|} + k_1(P_{xy:\overline{n}|} - P_{y:\overline{n}|}) + k_2(P_{xy:\overline{n}|} - P_{x:\overline{n}|}) \qquad (23·30)$$

or

$$P'_{xy:\overline{n}|} \doteqdot (1+k_1)\,P_{x:\overline{n}|} + (1+k_2)\,P_{y:\overline{n}|} - (1+k_1+k_2)\,P_{\overline{n}|}. \qquad (23·31)$$

23·11. Endowment assurance premiums at different rates of interest

Using the argument employed in section 8 to prove Lidstone's approximation but putting

$$\mu_t^{\alpha} = \mu_{x+t},$$

$$\mu_t^{\beta} = \delta' - \delta$$

and

$$\mu_t^{\gamma} = 0, \quad t < n,$$

$$= \infty, \quad t = n,$$

we have

$$\bar{P}'_{x:\overline{n}|} \doteqdot \bar{P}_{x:\overline{n}|} + \bar{P}'_{\overline{n}|} - \bar{P}_{\overline{n}|},$$

where unaccented functions are at rate of interest i (corresponding to the force δ) and accented functions are at rate i' (corresponding to the force δ').

Similarly

$$P'_{x:\overline{n}|} \doteqdot P_{x:\overline{n}|} + P'_{\overline{n}|} - P_{\overline{n}|}. \qquad (23·32)$$

By means of formula (23·32), $P'_{x:\overline{n}|}$ can be evaluated approximately if $P_{x:\overline{n}|}$ is known and if compound interest functions at rates i and i' are available.

Since formula (23·32) can be written in the form

$$P'_{x:\overline{n}|} - P_{x:\overline{n}|} \doteqdot P'_{\overline{n}|} - P_{\overline{n}|},$$

the right-hand side of which is independent of the mortality table, it follows that if we have two tables I and II

$$P'^{\,\mathrm{I}}_{x:\overline{n}|} - P^{\,\mathrm{I}}_{x:\overline{n}|} \doteqdot P'^{\,\mathrm{II}}_{x:\overline{n}|} - P^{\,\mathrm{II}}_{x:\overline{n}|}$$

or

$$P'^{\,\mathrm{I}}_{x:\overline{n}|} \doteqdot P^{\,\mathrm{I}}_{x:\overline{n}|} + P'^{\,\mathrm{II}}_{x:\overline{n}|} - P^{\,\mathrm{II}}_{x:\overline{n}|}, \qquad (23·33)$$

i.e. $P'^{\,\mathrm{I}}_{x:\overline{n}|}$ can be evaluated approximately if $P^{\,\mathrm{I}}_{x:\overline{n}|}$, $P^{\,\mathrm{II}}_{x:\overline{n}|}$ and $P'^{\,\mathrm{II}}_{x:\overline{n}|}$ are all available.

Formula (23·33) is more accurate than formula (23·32) because the errors tend to cancel out. In fact, formula (23·33) holds also for whole-life assurances, although with less accuracy than for endowment assurances. It forms the basis of Wickens's method (described in Volume I, section 9·3) whereby values of P^{II}_x at three rates of interest, and of P^{I}_x at two of those rates, are used in order to obtain a close approximation to P^{I}_x at the third rate.

*23·12. The calculation of endowment assurance premiums when no monetary functions are available

From formula (23·32),

$$P'_{x:\overline{n}|} \doteq P'_{\overline{n}|} + (P_{x:\overline{n}|} - P_{\overline{n}|}).$$

Putting $\delta = 0$ and writing δ for δ', we have

$$P_{x:\overline{n}|} \doteq P_{\overline{n}|} + \left(\frac{1}{\ddot{e}_{x:\overline{n}|}} - \frac{1}{n}\right). \tag{23·34}$$

This formula provides a means of evaluating $P_{x:\overline{n}|}$ approximately when no monetary functions based on the life table are available. Unlike formula (23·19), it almost always gives too small a value.

Another approximation can be obtained as follows. $\ddot{a}_{x:\overline{n}|}/\ddot{a}_{\overline{n}|}$ is the weighted mean of ${}_tp_x$ $(t = 0, 1, 2, \ldots, n-1)$ with weights $1, v, v^2, \ldots, v^{n-1}$. $\ddot{e}_{x:\overline{n}|}/n$ is the unweighted mean of the same series. Hence

$$\frac{\ddot{a}_{x:\overline{n}|}}{\ddot{a}_{\overline{n}|}} \doteq \frac{\ddot{e}_{x:\overline{n}|}}{n},$$

i.e.

$$\frac{1}{\ddot{a}_{x:\overline{n}|}} \doteq \frac{1}{\ddot{e}_{x:\overline{n}|}} \cdot \frac{n}{\ddot{a}_{\overline{n}|}}.$$

Now

$$P_{x:\overline{n}|} = P_{\overline{n}|} + \frac{1}{\ddot{a}_{x:\overline{n}|}} - \frac{1}{\ddot{a}_{\overline{n}|}}.$$

$$\therefore\quad P_{x:\overline{n}|} \doteq P_{\overline{n}|} + \left(\frac{1}{\ddot{e}_{x:\overline{n}|}} - \frac{1}{n}\right)\frac{n}{\ddot{a}_{\overline{n}|}}. \tag{23·35}$$

Since ${}_tp_x$ is a decreasing series,

$$\frac{\ddot{a}_{x:\overline{n}|}}{\ddot{a}_{\overline{n}|}} > \frac{\ddot{e}_{x:\overline{n}|}}{n}.$$

It follows that formula (23·35) gives too large a value.

By means of the approximate relationship

$$\frac{n}{\ddot{a}_{\overline{n}|}} \doteq 1 + \tfrac{1}{2}ni,$$

formula (23·35) can be converted into the form

$$P_{x:\overline{n}|} \doteq P_{\overline{n}|} + \left(\frac{1}{\ddot{e}_{x:\overline{n}|}} - \frac{1}{n}\right)(1 + \tfrac{1}{2}ni). \tag{23·36}$$

H. Jecklin (*Proc. Cent. Assembly* [1948], **2**, 174) has suggested the following formula, obtained by taking the mean of formulae (23·34) and (23·36)

$$P_{x:\overline{n}|} \doteq P_{\overline{n}|} + \left(\frac{1}{\ddot{e}_{x:\overline{n}|}} - \frac{1}{n}\right)(1 + \tfrac{1}{4}ni). \tag{23·37}$$

The following table gives examples of premiums calculated by this formula according to the A 1924–29 ultimate mortality table with interest at 3 per cent.

TABLE 23·1

x	n	$P_{\overline{n}\|}$	$\frac{1}{\ddot{e}_{x:\overline{n}\|}}$	$1+\frac{1}{4}ni$	$P_{x:\overline{n}\|}$ approx.	$P_{x:\overline{n}\|}$ true
30	10	·08469	·101193	1·0750	·08597	·08598
	15	·05220	·068029	1·1125	·05372	·05372
	20	·03613	·051553	1·1500	·03792	·03791
40	10	·08469	·102096	1·0750	·08694	·08694
	15	·05220	·069157	1·1125	·05497	·05496
	20	·03613	·052988	1·1500	·03957	·03952

EXAMPLES

Example 23·1

A 10-year temporary assurance of £1,000 is required in respect of four lives all aged 50. If only one of the lives dies during the term, the sum assured will be paid at the end of the term, but if two lives die, the sum assured will be paid at the end of the year in which the second death occurs. If no claim occurs, one-half of the premiums paid is to be returned.

Develop a formula for the annual premium payable throughout the duration of the assurance, assuming an expense loading of 10 per cent of the office premium.

Solution

Denote the annuity for valuing the premiums by $\ddot{a}$. This annuity will be payable as long as at least 3 lives survive with a maximum of ten years.

$$\frac{Z^3}{(1+Z)^3}=Z^3-3Z^4+\dots$$

$$\therefore \quad \ddot{a}=4\ddot{a}_{50:\,50:\,50:\,\overline{10}|}-3\ddot{a}_{50:\,50:\,50:\,50:\,\overline{10}|}.$$

If we use the single premium conversion relationship, $A=1-d\ddot{a}$, we shall obtain an assurance payable as follows:

If no death occurs—at the end of the term,

If only one life dies—at the end of the term,

If two or more lives die—at the end of the year of the second death.

It will be seen that this assurance exceeds the one described in the question by

$$E={}_{10}E_{50:\,50:\,50:\,50}.$$

Therefore, if we denote the required assurance by A,

$$A=1-d\ddot{a}-E.$$

Let P be the office premium. Then, equating net premiums to benefits, we have

$$\cdot 9P\ddot{a}=1000A+5PE.$$

$$\therefore \quad P=\frac{1000A}{\cdot 9\ddot{a}-5E}.$$

Example 23·2

A special life assurance policy for a life aged x provides, in return for a level annual premium, a sum of £1,000 plus one-half of the reserve under the policy at the end of the year of death whenever it may occur.

Show how to calculate the net annual premium for this policy, assuming that a special mortality table is available under which the rate of mortality at every age is one-half of the normal rate.

Solution

From formula (23·15) the annual premium (P) and the reserve (V_t) for the policy are related by the formula:

$$V_{t+1}-V_t=P_t+(V_t+P_t)\,i-\{(1000+\tfrac{1}{2}V_{t+1})-V_{t+1}\}\,q_t.$$

This may be written

$$V_{t+1}-V_t=P_t+(V_t+P_t)\,i-\{2000-V_{t+1}\}\frac{q_t}{2}.$$

Hence, the premium required is equal to the annual premium for an ordinary £2,000 whole-life assurance according to the special mortality table.

Example 23·3

An office decides to issue a 20-year joint-life endowment assurance on two impaired lives aged 40. It is considered that the impairments are equivalent to a 20 per cent addition to the force of mortality in each case. Calculate, by the method of section 23·10, the net extra premium required on the basis of the Hypothetical Select Mortality Table, ultimate mortality, with 2 per cent interest.

Compare your result with that obtained by assuming an approximate equivalent rating-up in age.

Solution

The net extra premium required, by the method of section 23·10, is

$$\cdot 2P_{40:\overline{20}|}+\cdot 2P_{40:\overline{20}|}-\cdot 4P_{\overline{20}|}=\cdot 4\left(\frac{1}{\ddot{a}_{40:\overline{20}|}}-d\right)-\cdot 4\left(\frac{1}{\ddot{a}_{\overline{20}|}}-d\right)$$

$$=\cdot 4\left(\frac{1}{\ddot{a}_{40:\overline{20}|}}-\frac{1}{\ddot{a}_{\overline{20}|}}\right).$$

Now $$\frac{1}{\ddot{a}_{40:\overline{20}|}}=\frac{D_{40}}{N_{40}-N_{60}}=\frac{214{,}984}{5{,}099{,}958-1{,}677{,}004}=\cdot 0628$$

and $$\frac{1}{\ddot{a}_{\overline{20}|}}=\frac{1}{16\cdot 6785}=\cdot 0600.$$

Hence the net extra premium is

$$\cdot 4\times\cdot 0028=\cdot 0011.$$

By rating up the age we have:

Age	μ_x	$1\cdot 2\mu_x$	Rated-up age
40	·00295	·00354	43
50	·00668	·00802	52
60	·01868	·02242	62

This suggests an addition of 2 years to age. The net extra premium will therefore be approximately

$$\left(\frac{1}{\ddot{a}_{42:42:\overline{20}|}}-\frac{1}{\ddot{a}_{40:40:\overline{20}|}}\right).$$

Now $$\frac{1}{\ddot{a}_{42:42:\overline{20}|}}=\frac{968{,}621}{18{,}543{,}774-4{,}054{,}294}=\cdot 0668$$

and $$\frac{1}{\ddot{a}_{40:40:\overline{20}|}}=\frac{1{,}020{,}513}{20{,}558{,}721-5{,}021{,}882}=\cdot 0657.$$

Hence the net extra premium is approximately $\cdot 0668-\cdot 0657=\cdot 0011$.

CHAPTER 24

CONTINGENT ASSURANCES

24·1. Introduction

A contingent assurance is an assurance on two or more lives under which the sum assured is payable only if the deaths occur in a specified order.

In this chapter it will be assumed that the probabilities of living and dying in respect of the individual lives are independent and hence that any required compound probability can be obtained by the multiplication rule.

The expressions for assurances will be developed on a continuous basis on the assumption that the sum assured is payable at the moment of death. The main reason why it is frequently assumed, in the case of single-life and joint-life assurances, that the sum assured is payable at the end of the year of death (an assumption which is not in accordance with office practice) is that, on this assumption, the premium conversion relationships between the annuity-value and the single and annual premiums hold good. As these relationships do not apply to contingent assurances and as the continuous method is particularly suitable, not only for developing expressions for complicated contingent assurances but also for certain approximate methods of evaluation which will be explained in Chapters 26 and 27, there is usually no advantage in making the assumption that the sum assured is payable at the end of the year of death. However, if expressions for assurances payable at the end of the year of death are required they may be obtained from those developed in this chapter by merely altering $\bar{A}$ to A.

24·2. Probabilities involving two lives

The symbol ${}_nq^1_{xy}$ is used to represent the probability that (x) will die before (y) and within n years. As $\delta t \to 0$, the probability that (x) and (y) will both survive t years and (x) will die in the short interval between time t and time $t+\delta t$ tends to

$$ {}_tp_{xy}\,\mu_{x+t}\,\delta t. $$

It follows that

$$ {}_nq^{1}_{xy} = \int_0^n {}_tp_{xy}\,\mu_{x+t}\,dt. \tag{24·1} $$

Putting $n=\infty$, we have, as the probability that (x) will die before (y),

$$ {}_\infty q^{1}_{xy} = \int_0^\omega {}_tp_{xy}\,\mu_{x+t}\,dt. \tag{24·2} $$

When $n=1$, the prefix is omitted and we write

$$ q^{1}_{xy} = \int_0^1 {}_tp_{xy}\,\mu_{x+t}\,dt. $$

These formulae may also be derived by an application of the principles of the multiple-decrement table, for from Chapter 19

$$ {}_n(aq)^{\alpha}_x = \int_0^n {}_t(ap)_x\,(a\mu)^{\alpha}_{x+t}\,dt, $$

and we have seen in Chapter 21 that the joint-life mortality table is a form of double-decrement table in which ${}_t(ap)_x$ is replaced by ${}_tp_{xy}$ and $(a\mu)^{\alpha}_{x+t}$ ($\equiv \mu^{\alpha}_{x+t}$) is replaced by μ_{x+t}. Hence formula (24·1) follows.

The symbol ${}_nq^{2}_{xy}$ is used to represent the probability that (x) will die after (y) and within n years. Using a similar line of reasoning to that by which formula (24·1) was obtained, we have

$$ \begin{aligned} {}_nq^{2}_{xy} &= \int_0^n (1-{}_tp_y)\,{}_tp_x\,\mu_{x+t}\,dt \\ &= \int_0^n {}_tp_x\,\mu_{x+t}\,dt - \int_0^n {}_tp_{xy}\,\mu_{x+t}\,dt. \end{aligned} $$

$$ \therefore\quad {}_nq^{2}_{xy} = {}_nq_x - {}_nq^{1}_{xy}. \tag{24·3} $$

This is obviously correct because if (x) dies within n years he must die either *before* (y) and within n years or *after* (y) and within n years. Putting $n=\infty$ in formula (24·3), we have

$$ {}_\infty q^{2}_{xy} = 1 - {}_\infty q^{1}_{xy}, \tag{24·4} $$

which again is obviously correct. Also, the probability that (x) will die second of the two lives is the same as the probability that (y) will die first.

$$ \therefore\quad {}_\infty q^{2}_{xy} = {}_\infty q_{x\overset{1}{y}}. $$

It should be noted, however, that

$$ {}_nq^{2}_{xy} \neq {}_nq_{x\overset{1}{y}}. $$

24·3. Contingent assurances involving two lives

The symbol $\bar{A}^{1}_{xy}$ is used to represent the value of an assurance of 1 payable on the death of (x) if he dies before (y). (x) is called the life assured or the failing life and (y) the counter life.

$$\bar{A}^{1}_{xy} = \int_0^\infty v^t\,{}_tp_{xy}\,\mu_{x+t}\,dt. \tag{24·5}$$

It will be seen that $\bar{A}^{1}_{xy} + \bar{A}_{x\overset{1}{y}} = \bar{A}_{xy}$

and hence that $\bar{A}^{1}_{xx} = \tfrac{1}{2}\bar{A}_{xx}$.*

The value of an assurance of 1 payable on the death of (x) if he dies after (y) is denoted by the symbol $\bar{A}^{2}_{xy}$ in which the 2 is placed over the x in order to indicate not only that (x) must die second but that the sum assured is payable on the second death.

$$\left.\begin{aligned}\bar{A}^{2}_{xy} &= \int_0^\infty v^t(1 - {}_tp_y)\,{}_tp_x\,\mu_{x+t}\,dt\\ &= \bar{A}_x - \bar{A}^{1}_{xy}.\end{aligned}\right\} \tag{24·6}$$

The latter expression is obviously correct since the combination of an assurance payable on the death of (x) if he dies *after* (y) and an assurance payable on the death of (x) if he dies *before* (y) is an assurance payable on the death of (x) whenever he may die.

Since $\bar{A}^{2}_{xy} = \bar{A}_x - \bar{A}^{1}_{xy}$

and $\bar{A}_{x\overset{2}{y}} = \bar{A}_y - \bar{A}_{x\overset{1}{y}}$,

it follows that $\bar{A}^{2}_{xy} + \bar{A}_{x\overset{2}{y}} = \bar{A}_x + \bar{A}_y - (\bar{A}^{1}_{xy} + \bar{A}_{x\overset{1}{y}})$,

i.e. $\bar{A}_{\overline{xy}} = \bar{A}_x + \bar{A}_y - \bar{A}_{xy}$,

which is formula (22·13).

24·4. Contingent assurances involving three lives

If we substitute the joint status yz for the lifetime of (y) in formula (24·5) we have an assurance payable on the death of (x) if he dies first of the three lives (x), (y) and (z).

$$\bar{A}^{1}_{xyz} = \int_0^\infty v^t\,{}_tp_{xyz}\,\mu_{x+t}\,dt. \tag{24·7}$$

* In this book, when the letter x occurs more than once in the same suffix, it is understood that the lives are not only of the same age but taken at random from the same group of lives.

It will be seen that

$$\bar{A}^{1}_{xyz}+\bar{A}_{x\overset{1}{y}z}+\bar{A}_{xy\overset{1}{z}}=\bar{A}_{xyz}$$

and hence that $$\bar{A}^{1}_{xxx}=\tfrac{1}{3}\bar{A}_{xxx}.$$

Other assurances involving three lives can be expressed in integral form and most of the integrals can be converted into expressions consisting entirely of assurances depending on the first death; these expressions may be verified by general reasoning. For example

$$\left.\begin{aligned}\bar{A}^{3}_{xyz}&=\int_0^\infty v^t(1-{}_tp_y)(1-{}_tp_z){}_tp_x\,\mu_{x+t}\,dt\\&=\bar{A}_x-\bar{A}^{1}_{xy}-\bar{A}^{1}_{xz}+\bar{A}^{1}_{xyz},\end{aligned}\right\}\qquad(24\cdot8)$$

and

$$\left.\begin{aligned}\bar{A}^{1}_{x:\overline{yz}}&=\int_0^\infty v^t\,{}_tp_x\cdot{}_tp_{\overline{yz}}\,\mu_{x+t}\,dt\\&=\int_0^\infty v^t\,{}_tp_x({}_tp_y+{}_tp_z-{}_tp_{yz})\,\mu_{x+t}\,dt\\&=\bar{A}^{1}_{xy}+\bar{A}^{1}_{xz}-\bar{A}^{1}_{xyz}.\end{aligned}\right\}\qquad(24\cdot9)$$

The assurance represented by formula (24·9) is payable on the death of (x) if he dies either first or second of the three lives. It follows that

$$\left.\begin{aligned}\bar{A}^{2}_{xyz}&=\bar{A}^{1}_{x:\overline{yz}}-\bar{A}^{1}_{xyz}\\&=\bar{A}^{1}_{xy}+\bar{A}^{1}_{xz}-2\bar{A}^{1}_{xyz}.\end{aligned}\right\}\qquad(24\cdot10)$$

An alternative symbol for $\bar{A}^{1}_{x:\overline{yz}}$ is $\bar{A}^{1:2}_{x:y:z}$.

The value of an assurance of 1 payable on the failure of the joint status xy, provided that this occurs before the death of (z), is denoted by the symbol $\bar{A}^{\;\;1}_{\overline{xy}:z}$.

$$\left.\begin{aligned}\bar{A}^{\;\;1}_{\overline{xy}:z}&=\int_0^\infty v^t\,{}_tp_{xyz}(\mu_{x+t}+\mu_{y+t})\,dt\\&=\bar{A}^{1}_{xyz}+\bar{A}_{x\overset{1}{y}z}.\end{aligned}\right\}\qquad(24\cdot11)$$

In the assurances so far discussed in this section, the order of the deaths is not fully specified. For example in $\bar{A}^{2}_{xyz}$ it is immaterial whether (y) or (z) dies first. There are two types of assurance, involving three lives, in which the order of the deaths is fully specified. These are (i) an assurance payable on the *second* death provided that a specified life dies first and (ii) an assurance payable on the *third* death provided that a specified life dies first. They are typified by the symbols $\bar{A}^{\,2}_{x\underset{1}{y}z}$ and $\bar{A}^{\,3}_{x\underset{1}{y}z}$ respectively. In these symbols the figure 1 is placed under the y in order to indicate that (y) must die first but that no payment is to be made on his death; the figure 2 or 3 is placed over the x in order to indicate not only that (x) must die second or third (as the case may be) but that the sum assured is payable on his death.

$$\left.\begin{aligned}\bar{A}^{\,2}_{x\underset{1}{y}z} &= \int_0^\infty v^t(1-{}_tp_y)\,{}_tp_{xz}\,\mu_{x+t}\,dt\\ &= \bar{A}^{\,1}_{xz} - \bar{A}^{\,1}_{xyz}.\end{aligned}\right\} \tag{24·12}$$

It will be seen that

$$\begin{aligned}\bar{A}^{\,2}_{xyz} &= \bar{A}^{\,2}_{x\underset{1}{y}z} + \bar{A}^{\,2}_{xy\underset{1}{z}}\\ &= \bar{A}^{\,1}_{xy} + \bar{A}^{\,1}_{xz} - 2\bar{A}^{\,1}_{xyz},\end{aligned}$$

confirming formula (24·10). It will also be seen that

$$\bar{A}^{\,1}_{xyz} + \bar{A}^{\,2}_{xyz} + \bar{A}^{\,3}_{xyz} = \bar{A}_x,$$

which is obviously correct.

In all the expressions in integral form which have so far been written down in respect of contingent assurances, the integration has been based on the happening of the event on which the sum assured is payable. If we follow this procedure with $\bar{A}^{\,3}_{x\underset{1}{y}z}$, we have

$$\bar{A}^{\,3}_{x\underset{1}{y}z} = \int_0^\infty v^t\,{}_tq^{\;2}_{\underset{1}{y}z}\cdot{}_tp_x\,\mu_{x+t}\,dt.$$

This expression is not in a very satisfactory form, since it contains the term ${}_tq^{\;2}_{yz}$. A more useful expression is obtained by basing the integration on the second death, bearing in mind that on the death of (z), (x) being alive and (y) having died previously, the assurance becomes a straightforward assurance on the life of (x). By this method we have

$$\bar{A}^{\,3}_{x\underset{1}{y}z} = \int_0^\infty v^t(1-{}_tp_y)\,{}_tp_{xz}\,\mu_{z+t}\,\bar{A}_{x+t}\,dt. \tag{24·13}$$

This expression cannot be converted into one consisting entirely of assurances payable on the first death. However, it is in a more convenient form for evaluation than the expression previously obtained (see Chapter 26).

A contingent assurance which is payable on the failure of a last survivor status cannot be expressed in the form of a single integral because, as pointed out in Chapter 22, the force of decrement is not predetermined. For example, the value of an assurance payable on the death of the survivor of (x) and (y) should that occur before the death of (z), is

$$\bar{A}^{\,2}_{\underset{1}{x}yz}+\bar{A}_{\underset{1}{x}y\overset{2}{z}}=\bar{A}_{\overset{1}{x}z}+\bar{A}_{\overset{1}{y}z}-\bar{A}_{\overset{1}{x}yz}-\bar{A}_{x\overset{1}{y}z}.$$

24·5. Contingent assurances involving more than three lives

Expressions for contingent assurances involving more than three lives can be written down in integral form. The simplest of these assurances is

$$\bar{A}_{\overset{1}{x}yzw}=\int_0^\infty v^t\,{}_tp_{xyzw}\,\mu_{x+t}\,dt.$$

Some of the more complicated assurances can be expressed completely in terms of assurances depending on the first death. For example

$$\begin{aligned}\bar{A}^{\,3}_{xyzw}&=\int_0^\infty v^t\,{}_tp_x\cdot{}_tp^{[1]}_{\overline{yzw}}\,\mu_{x+t}\,dt\\&=\int_0^\infty v^t\,{}_tp_x({}_tp_y+{}_tp_z+{}_tp_w-2\,{}_tp_{yz}-2\,{}_tp_{yw}\\&\qquad\qquad-2\,{}_tp_{zw}+3\,{}_tp_{yzw})\,\mu_{x+t}\,dt\\&=\bar{A}_{\overset{1}{x}y}+\bar{A}_{\overset{1}{x}z}+\bar{A}_{\overset{1}{x}w}-2\bar{A}_{\overset{1}{x}yz}-2\bar{A}_{\overset{1}{x}yw}-2\bar{A}_{\overset{1}{x}zw}+3\bar{A}_{\overset{1}{x}yzw}.\end{aligned}$$

However, there is usually little or no advantage, in practice, in being able to express a complicated assurance in this way.

We have already seen that when three lives are involved it is best to base the integration on the death which results in payment of the sum assured, except in the case of $\bar{A}^{\,3}_{\underset{1}{x}yz}$ for which an integral based on the second death gives the simplest expression. When more than three lives are involved it is best to choose the latest death which does not necessitate the inclusion of a contingent probability in the expression to be integrated. For example,

$$\bar{A}^{\,3}_{\underset{12}{xyzw}}=\int_0^\infty v^t(1-{}_tp_y)\,{}_tp_{xzw}\,\mu_{z+t}\,\bar{A}_{\overset{1}{x+t}:w+t}\,dt.$$

In this case an integration based on the death of (x) would necessitate the inclusion of the contingent probability ${}_tq^{\,1}_{yz}$; generally speaking an expression containing $\bar{A}_{x+t:w+t}^{\,1}$ is more convenient for evaluation than one containing ${}_tq^{\,1}_{yz}$.

Complicated contingent assurances involving more than three lives rarely occur in practice.

24·6. Temporary contingent assurances

An expression for the value of a temporary contingent assurance can be obtained by altering the limits of integration. Provided that the assurance is payable on the first death, the value can also be obtained by deducting the value of a deferred contingent assurance from the value of the whole-life contingent assurance. Thus

$$
\begin{aligned}
{}_n\bar{A}^{\,1}_{xy} &= \int_0^n v^t\,{}_tp_{xy}\,\mu_{x+t}\,dt \\
&= \bar{A}^{\,1}_{xy} - v^n\,{}_np_{xy}\,\bar{A}^{\,1}_{x+n:\,y+n}.
\end{aligned}
$$

24·7. Selection

The formulae developed in this chapter can easily be extended to select tables. For example,

$$\bar{A}^{\,1}_{[x]:\,[y]} = \int_0^\infty v^t\,{}_tp_{[xy]}\,\mu_{[x]+t}\,dt.$$

24·8. Annual premiums. Medical examination

For a simple contingent assurance payable on the death of (x) if he predeceases (y), the annual premiums should continue only during the joint lifetime of (x) and (y), i.e. the net annual premium should be $\bar{A}^{\,1}_{xy}/\ddot{a}_{xy}$. A similar principle applies when more than two lives are involved and the sum assured is payable on the death of (x) if he dies first. For example, the net annual premium for the assurance $\bar{A}^{\,1}_{xyz}$ is $\bar{A}^{\,1}_{xyz}/\ddot{a}_{xyz}$. In all such cases it is necessary to have some evidence that the life assured is a first-class life; for this purpose it is customary to have him medically examined. There is no need to examine the counter life or lives because the death of any counter life causes the assurance to terminate.

If the sum assured is not payable on the first death, the status during which the premiums are to be payable needs careful consideration. It is also necessary to decide which of the lives shall be medically examined. As this depends partly on the status of payment of the premiums it is proposed to consider the two problems together.

For the assurance $\bar{A}^{2}_{xy}$ the premiums should be made payable during the lifetime of (x), i.e. the net premium should be $\bar{A}^{2}_{xy}/\ddot{a}_x$, because the assurance will terminate on the death of (x) but not on the death of (y). It is necessary to examine (y) because his early death will increase the value of the policy. It is usually necessary to examine (x) also. Care is necessary, however, because an addition to the age of (x)—a common method of dealing with impaired lives assured under contingent assurances—may reduce the premium $\bar{A}^{2}_{xy}/\ddot{a}_x$ instead of increasing it. This is particularly likely to happen if (x) is much older than (y). If, therefore, the office is content to obtain its extra premiums for contingent assurances by merely increasing the age of the life assured and if, in any individual case, tests show that $P''^{\;2}_{x+n:y}$ is less than P''^{2}_{xy} for all values of n, there is no point in examining (x). (It is assumed that the office is not prepared to quote a reduced premium should the examination show that (x) is an impaired life.)

In more complicated cases the main principles to be borne in mind are

(i) that, as the assured cannot be expected to pay premiums in respect of a policy which is no longer in force, the premiums must cease on any death which causes the assurance to terminate,

(ii) that it is necessary to examine anyone whose early death will improve the value of the policy, either by increasing the probability that the sum assured will ultimately become payable or by making the assurance paid-up,

(iii) that it is usually necessary to examine the person at whose death the sum assured is payable.

These principles may be illustrated by a consideration of the assurance $\bar{A}^{2}_{x\underset{1}{y}z}$. The net premium for this assurance should be $\bar{A}^{2}_{x\underset{1}{y}z}/\ddot{a}_{xz}$ because the assurance will terminate as soon as either (x) or (z) dies but it will not terminate on the death of (y). It is necessary to examine (y), because his early death increases the probability that the sum assured will ultimately become payable. It is also necessary to examine (x), unless the effect of rating him up is to reduce the premium. There is no point in examining (z), because his death during the currency of the assurance causes it to terminate.

24·9. The possibility of an option against the office

With some types of contingent assurance there is a danger that, on the death of one of the lives, the assured will have the option of dropping the policy and effecting another one which provides the same benefit at a lower annual premium. An example of this is given by the assurance $\bar{A}^1_{x:\overline{yz}}$. As this assurance will terminate on the failure of the joint status $x:\overline{yz}$ it seems at first sight that a premium continuing during this status should be quoted. If this is done, however, and if (y) or (z) dies during the tth year, the assured will have the option, if (x) is in good health and if the premium $P''^{\ 1}_{x+t:z+t}$ or $P''^{\ 1}_{x+t:y+t}$ (as the case may be) is lower than the original premium, of dropping the policy and effecting a new one at this lower premium, thus involving the office in a loss. The risk of this cannot be eliminated by examining (y) and (z) because mere examination will not prevent their early death.

One way out of the difficulty is to charge a premium which is payable during the joint lifetime of (x), (y) and (z). This is open to two objections—first, that it increases the premium and therefore may make it unattractive to the assured, and secondly, that it necessitates examination of all three lives.

Another method is to charge a premium which is payable during the status $x:\overline{yz}$ but which, instead of being level throughout this status, is larger during the joint lifetime of (x), (y) and (z) and reduces on the death of (y) or (z). Thus if P'' denotes the level office premium payable during the status $x:\overline{yz}$ and if it is desired to charge instead a premium of ${}^1P''$ reducing to ${}^2P''$ on the death of (y) or (z), ${}^2P''$ can be chosen so that it is equal to the smaller of $P''^{\ 1}_{x+1:y+1}$ and $P''^{\ 1}_{x+1:z+1}$ and ${}^1P''$ can be obtained from the equation of value

$$P''\ddot{a}_{x:\overline{yz}} = {}^2P''\ddot{a}_{x:\overline{yz}} + ({}^1P'' - {}^2P'')\,\ddot{a}_{xyz}.$$

(It is assumed that, as is usually the case, $P''^{\ 1}_{x+t:y+t}$ and $P''^{\ 1}_{x+t:z+t}$ both increase as t increases.)

If this method is adopted and the ages of (y) and (z) are not approximately equal, it may be considered advisable to obtain some evidence of health of the older of them (not necessarily by medical examination) because his early death would convert the policy into a simple contingent assurance at an inadequate premium.

A practical point to be borne in mind is that these policies are usually effected in connexion with legal transactions so that, even if the assured realizes that he has the option of exchanging the policy for another one at a lower premium, the expense involved may make the change unattractive. The office may therefore decide to charge a level premium during the status $x:\overline{yz}$, in spite of the option. However, the possibility that the option may arise must not be overlooked.

24·10. Office premiums

Office premiums for contingent assurances are usually obtained by adding a percentage and a constant loading to the select net premium, the loadings being similar to those used for temporary assurances. If the net annual premium is very small the office annual premium obtained by adding the usual loadings may not be sufficient to cover the expenses. In such a case it is necessary either to increase the office annual premium to an arbitrary figure or to quote a single premium. If the risk is a remote one a precise calculation of the net premium will not be necessary; a rough estimate of the net single premium will be made and an arbitrary office single premium sufficient to cover the risk and expenses will be quoted.

24·11. Policy values

The value, at the end of t years, of a contingent assurance by annual premiums assuring 1 on the death of (x) before (y) is

$$\bar{A}^{\,1}_{x+t:y+t} - P\ddot{a}_{x+t:y+t},$$

where

$$P = \bar{A}^{\,1}_{xy}/\ddot{a}_{xy}.$$

If the sum assured is payable on the death of (x) after (y), the value depends on whether both (x) and (y) are alive or only (x) is alive. In the former case the value is

$$\bar{A}^{\,2}_{x+t:y+t} - P\ddot{a}_{x+t}$$

and in the latter it is $\bar{A}_{x+t} - P\ddot{a}_{x+t}$,

where $P = \bar{A}^2_{xy}/\ddot{a}_x$.

Expressions for the values of other contingent assurances can be obtained similarly.

24·12. Evaluation of contingent assurances

The problem of calculating arithmetical values of contingent assurances will be investigated in Chapters 26 and 27.

EXAMPLES

Example 24·1

Find an expression for the annual premium for an assurance payable if (x) dies before (y) or within 10 years after (y).

Solution

The value of the assurance is

$$\bar{A}^{1}_{x:\overline{10|}} + \int_{10}^{\infty} v^t \, {}_{t-10}p_y \, {}_tp_x \, \mu_{x+t} dt$$

$$= \bar{A}^{1}_{x:\overline{10|}} + \int_{0}^{\infty} v^{10+t} \, {}_tp_y \, {}_{10+t}p_x \, \mu_{x+10+t} dt$$

$$= \bar{A}^{1}_{x:\overline{10|}} + {}_{10}E_x \bar{A}^{\;\;1}_{x+10:y}.$$

The annual premium may be made payable during the continuance of the assurance, the annuity-value being

$$\ddot{a}_{x:\overline{10|}} + {}_{10}E_x \ddot{a}_{x+10:y}.$$

If this is done and if (x) outlives (y), the last annual premium will fall due when the unexpired term of the assurance is less than one year. The assured may decide not to pay this last premium, in which event, if the reserve for the final fraction of a year's assurance is less than the net premium, the office will suffer a small theoretical loss.

One way out of the difficulty is to make the premium payable during the joint lifetime of the two lives. This is open to two objections—first that it increases the premium, and therefore may make it unattractive to the assured, and secondly that it necessitates medical examination of (y).

Probably the best plan is to arrange for only nine premium payments after the death of (y), i.e. to use the annuity-value

$$\ddot{a}_{x:\overline{9|}} + {}_9E_x\,\ddot{a}_{x+9:y}.$$

Example 24·2

Find by the English Life Table No. 10, Males, the probability that three lives aged 40, 30 and 20 respectively will die in that order, at least ten years separating any two deaths.

Solution

$$\int_{10}^{\infty}(1-{}_{t-10}p_{40})\,{}_tp_{30}\,\mu_{30+t}\cdot{}_{10+t}p_{20}\,dt$$

$$=\int_0^{\infty}(1-{}_tp_{40})\,{}_{10+t}p_{30}\,\mu_{40+t}\cdot{}_{20+t}p_{20}\,dt$$

$$={}_{10}p_{30}\cdot{}_{20}p_{20}\int_0^{\infty}(1-{}_tp_{40})\,{}_tp_{40}\,\mu_{40+t}\cdot{}_tp_{40}\,dt$$

$$={}_{10}p_{30}\cdot{}_{20}p_{20}\left({}_{\infty}q^{\;1}_{40:40}-{}_{\infty}q^{\;1}_{40:40:40}\right)$$

$$=\frac{l_{40}}{l_{30}}\cdot\frac{l_{40}}{l_{20}}\left(\tfrac{1}{2}-\tfrac{1}{3}\right)=\cdot148.$$

Rough check

$$\mathring{e}_{30}=38.$$

$$(1-{}_{28}p_{40})\,{}_{48}p_{20}=\cdot22.$$

This does not agree very well, but of course the assumption that if the deaths occur in the prescribed order (30) will necessarily die at the end of his expectation is not a good one.

Example 24·3

Express, in terms of contingent assurances payable on the first death, the value of a contingent assurance payable on the death of the survivor of (20) and (30) provided that he attains at least age 60 and dies at least 10 years after the death of (40). You should assume that all three lives are taken at random from the same group of lives.

Solution

The value of the contingent assurance is

$$\int_{40}^{\infty} v^t \, {}_tp_{20}(1 - {}_tp_{30})(1 - {}_{t-10}p_{40})\, \mu_{20+t}\, dt$$
$$+ \int_{30}^{\infty} v^t \, {}_tp_{30}(1 - {}_tp_{20})(1 - {}_{t-10}p_{40})\, \mu_{30+t}\, dt$$

$$= v^{40} \, {}_{40}p_{20} \left[\int_0^{\infty} v^t \, {}_tp_{60}\, \mu_{60+t}\, dt - {}_{40}p_{30} \int_0^{\infty} v^t \, {}_tp_{60:70}\, \mu_{60+t}\, dt \right.$$
$$\left. - {}_{30}p_{40} \int_0^{\infty} v^t \, {}_tp_{60:70}\, \mu_{60+t}\, dt + {}_{40}p_{30} \cdot {}_{30}p_{40} \int_0^{\infty} v^t \, {}_tp_{60:70:70}\, \mu_{60+t}\, dt \right]$$
$$+ v^{30} \, {}_{30}p_{30} \left[\int_0^{\infty} v^t \, {}_tp_{60}\, \mu_{60+t}\, dt - {}_{30}p_{20} \int_0^{\infty} v^t \, {}_tp_{50:60}\, \mu_{60+t}\, dt \right.$$
$$\left. - {}_{20}p_{40} \int_0^{\infty} v^t \, {}_tp_{60:60}\, \mu_{60+t}\, dt + {}_{30}p_{20} \cdot {}_{20}p_{40} \int_0^{\infty} v^t \, {}_tp_{50:60:60}\, \mu_{60+t}\, dt \right]$$

$$= A_{20:\overset{1}{\overline{40|}}} [\bar{A}_{60} - {}_{40}p_{30} \bar{A}^{1}_{60:70} - {}_{30}p_{40} \bar{A}^{1}_{60:70} + {}_{40}p_{30} \cdot {}_{30}p_{40} \bar{A}^{1}_{60:70:70}]$$
$$+ A_{30:\overset{1}{\overline{30|}}} [\bar{A}_{60} - {}_{30}p_{20} \bar{A}^{1}_{60:50} - \tfrac{1}{2} \cdot {}_{20}p_{40} \bar{A}_{60:60} + {}_{30}p_{20} \cdot {}_{20}p_{40} \bar{A}^{1}_{60:50:60}].$$

CHAPTER 25

REVERSIONARY ANNUITIES

25·1. Reversionary annuities payable continuously

An annuity which commences on the failure of a given status and continues thereafter during the remainder of another given status is known as a reversionary annuity.

The simplest form of reversionary annuity is one which commences on the death of (x), if (y) is then living, and continues thereafter during the remainder of the lifetime of (y). The value of such an annuity, if it is payable continuously at the rate of 1 per annum, is denoted by the symbol $\bar{a}_{x|y}$. It can be expressed in integral form in either of two ways:

(*a*) If (x) dies at time t and (y) is then alive an annuity which will have a value of $\bar{a}_{y+t}$ will become payable.

$$\therefore \quad \bar{a}_{x|y} = \int_0^\infty v^t {}_tp_{xy}\, \mu_{x+t} \bar{a}_{y+t}\, dt. \tag{25·1}$$

(*b*) A payment will be made at time t if (y) is then alive and (x) has died previously.

$$\therefore \quad \bar{a}_{x|y} = \int_0^\infty v^t (1 - {}_tp_x)\, {}_tp_y\, dt, \tag{25·2}$$

whence

$$\bar{a}_{x|y} = \bar{a}_y - \bar{a}_{xy}. \tag{25·3}$$

Formula (25·3) is obviously correct, from general reasoning.

The equivalence of formulae (25·1) and (25·2) can be shown as follows:

$$\begin{aligned}
\bar{a}_{x|y} &= \int_0^\infty v^t {}_tp_{xy}\, \mu_{x+t} \bar{a}_{y+t}\, dt \\
&= \int_0^\infty {}_tp_x\, \mu_{x+t} \int_t^\infty v^r {}_rp_y\, dr\, dt \\
&= \int_0^\infty v^t {}_tp_y \int_0^t {}_rp_x\, \mu_{x+r}\, dr\, dt \\
&= \int_0^\infty v^t {}_tp_y (1 - {}_tp_x)\, dt.
\end{aligned}$$

Formula (25·3) can be written

$$\bar{a}_{x|y} = \{a_y + \tfrac{1}{2} - \tfrac{1}{12}(\mu_y + \delta)\} - \{a_{xy} + \tfrac{1}{2} - \tfrac{1}{12}(\mu_x + \mu_y + \delta)\}$$

or

$$\bar{a}_{x|y} = a_y - a_{xy} + \tfrac{1}{12}\mu_x. \tag{25·4}$$

The last term is usually so small that it can be neglected in practical work. Hence

$$\bar{a}_{x|y} \doteqdot a_y - a_{xy}. \tag{25·5}$$

25·2. Reversionary annuities payable yearly or *m* times a year

In practice annuities cannot be paid continuously and are usually paid yearly or *m* times a year, *m* being an integer. There are three different types of annuity payable *m* times a year to (*y*) after the death of (*x*):

(i) An annuity which is set up immediately but in respect of which (*y*) is entitled only to those payments which fall due after the death of (*x*). The payments of $1/m$ therefore fall due at intervals of one *m*th of a year calculated from the present time and the first payment of reversionary annuity is a full payment of $1/m$ at the end of the *m*th of a year in which (*x*) dies.

(ii) An annuity which is similar to (i) except that the first payment received by (*y*) is a proportionate one calculated in respect of the period from the date of death of (*x*) to the end of the *m*th of a year in which he dies.

(iii) An annuity which is entered upon immediately on the death of (*x*) so that the successive payments of $1/m$ fall due at intervals of one *m*th of a year calculated from the date of his death. This is the usual type of reversionary annuity issued by a life office.

In each case the annuity may be curtate (i.e. ceasing with the last payment of $1/m$ falling due before the date of death of (*y*)) or complete (i.e. ceasing with a proportionate payment at the death of (*y*)).

An annuity of type (i) is represented by the symbol $a_{x|y}^{(m)}$ if it is curtate or $\mathring{a}_{x|y}^{(m)}$ if it is complete. When $m = 1$ the affix (m) is omitted. There is no recognized symbol for an annuity of type (ii). When the annuity is of type (iii) the sign $\wedge$ is added over the main symbol 'a' and we write $\hat{a}_{x|y}^{(m)}$ for a curtate annuity or $\hat{\mathring{a}}_{x|y}^{(m)}$ for a complete annuity.

In practice, it is customary to take the value of a reversionary annuity of 1 to (y) after (x) as being $a_y - a_{xy}$, whether the annuity is of type (i), (ii) or (iii), whether it is curtate or complete, and whatever may be the value of m. Sometimes, however, it is necessary to use a more accurate expression. For example, it may be desired to calculate the cost of converting an annuity from type (iii) to type (i) or of making a curtate annuity complete. The values of the various types will therefore be investigated in the next section.

It should be noted that the expression $a_y - a_{xy}$ is not usually sufficiently accurate when one of the lives is replaced by a term certain (unless $m = 1$ and the annuity is curtate, in which event the expression is accurate). For example

$$a^{(m)}_{\overline{n}|\,|x} = a^{(m)}_x - a^{(m)}_{x:\overline{n}|} \doteq \left(a_x + \frac{m-1}{2m}\right) - \left\{a_{x:\overline{n}|} + \frac{m-1}{2m}\left(1 - \frac{D_{x+n}}{D_x}\right)\right\}$$

$$\text{i.e. } a^{(m)}_{\overline{n}|\,|x} \doteq a_x - a_{x:\overline{n}|} + \frac{m-1}{2m}\cdot\frac{D_{x+n}}{D_x}.$$

The last term in this expression cannot usually be ignored. The expression just obtained is equal to

$$\frac{D_{x+n}}{D_x}\left(a_{x+n} + \frac{m-1}{2m}\right),$$

the usual expression for a deferred annuity to (x) payable m times a year.

25·3. More accurate expressions for annuities payable yearly or *m* times a year

A curtate annuity of type (i) is equal to the difference between an immediate annuity to (y) and an immediate joint-life annuity to (x) and (y). Hence

$$a^{(m)}_{x|y} = a^{(m)}_y - a^{(m)}_{xy} = a_y - a_{xy} + \frac{m^2-1}{12m^2}\mu_x. \tag{25·6}$$

The last term is usually so small that, even when it is desired to distinguish between the different types, the formula

$$a^{(m)}_{x|y} \doteq a_y - a_{xy} \tag{25·7}$$

may be regarded as sufficiently accurate.

The value of the corresponding complete annuity may be obtained by adding $\frac{1}{2m}\bar{A}_{xy}^{\;2}$. If, therefore, formula (25·7) is used for the curtate annuity, it follows that

$$\mathring{a}_{x|y}^{(m)} \doteq a_y - a_{xy} + \frac{1}{2m}\bar{A}_{xy}^{\;2}$$

or

$$\mathring{a}_{x|y}^{(m)} \doteq a_y - a_{xy} + \frac{1}{2m}(\bar{A}_y - \bar{A}_{xy}^{\;1}). \tag{25·8}$$

To obtain the value of a curtate annuity of type (ii) we may deduct from the right-hand side of formula (25·7) the value of a payment of $\frac{1}{2m}$ at the end of the mth of a year in which (x) dies. This may conveniently be taken as $\frac{1}{2m}\bar{A}_{xy}^1$. We therefore have as the value of a curtate annuity of type (ii)

$$a_y - a_{xy} - \frac{1}{2m}\bar{A}_{xy}^1. \tag{25·9}$$

The value of a complete annuity of type (ii) may be obtained by deducting $\frac{1}{2m}\bar{A}_{xy}^1$ from the right-hand side of formula (25·8) or by adding $\frac{1}{2m}\bar{A}_{xy}^{\;2}$ to formula (25·9). Either method gives

$$a_y - a_{xy} - \frac{1}{2m}(\bar{A}_{xy}^1 - \bar{A}_y + \bar{A}_{xy}^{\;1})$$

$$= a_y - a_{xy} - \frac{1}{2m}(\bar{A}_{xy} - \bar{A}_y). \tag{25·10}$$

Another method of obtaining an expression for the value of a complete annuity of type (ii) is as follows:

The value of the annuity is almost exactly represented by the expression $\mathring{a}_y^{(m)} - \mathring{a}_{xy}^{(m)}$. (If the proportionate payment to the date of the first death under the joint-life annuity $\mathring{a}_{xy}^{(m)}$ were to be made at the end of the mth of a year in which the first death occurs instead of at the moment of death the value of a complete annuity of type (ii) would be exactly equal to $\mathring{a}_y^{(m)} - \mathring{a}_{xy}^{(m)}$. The error involved in the use of this expression is therefore exceedingly small.) It follows that the value may be taken as

$$\left(a_y + \frac{m-1}{2m} + \frac{1}{2m}\bar{A}_y\right) - \left(a_{xy} + \frac{m-1}{2m} + \frac{1}{2m}\bar{A}_{xy}\right)$$

$$= a_y - a_{xy} - \frac{1}{2m}(\bar{A}_{xy} - \bar{A}_y),$$

which is formula (25·10). Alternatively, the value may be taken as

$$(a_y+\tfrac{1}{2})\left(1-\frac{\delta}{2m}\right)-(a_{xy}+\tfrac{1}{2})\left(1-\frac{\delta}{2m}\right)$$
$$=(a_y-a_{xy})\left(1-\frac{\delta}{2m}\right). \tag{25·11}$$

For a curtate annuity of type (iii), we have

$$a_{x|y}^{(m)}=\int_0^\infty v^t\,{}_tp_{xy}\,\mu_{x+t}\,a_{y+t}^{(m)}\,dt$$
$$\doteq\int_0^\infty v^t\,{}_tp_{xy}\,\mu_{x+t}\left(\bar{a}_{y+t}-\frac{1}{2m}\right)dt$$
$$\doteq a_y-a_{xy}-\frac{1}{2m}\bar{A}^{1}_{xy}. \tag{25·12}$$

To obtain the value of a complete annuity of type (iii) we may write

$$\mathring{a}_{x|y}^{(m)}=\int_0^\infty v^t\,{}_tp_{xy}\,\mu_{x+t}\,\mathring{a}_{y+t}^{(m)}\,dt.$$

Replacing $\mathring{a}_{y+t}^{(m)}$ by $\bar{a}_{y+t}-\dfrac{1}{2m}+\dfrac{1}{2m}\bar{A}_{y+t}$, we have

$$\mathring{a}_{x|y}^{(m)}\doteq a_y-a_{xy}-\frac{1}{2m}(\bar{A}_{xy}-\bar{A}_y). \tag{25·13}$$

Alternatively, replacing $\mathring{a}_{y+t}^{(m)}$ by $\bar{a}_{y+t}\left(1-\dfrac{\delta}{2m}\right)$, we have

$$\mathring{a}_{x|y}^{(m)}\doteq(a_y-a_{xy})\left(1-\frac{\delta}{2m}\right). \tag{25·14}$$

Comparing formulae (25·12), (25·13) and (25·14) with formulae (25·9), (25·10) and (25·11) respectively, we see that for practical purposes it makes no difference to the value of a reversionary annuity whether the annuity is of type (ii) or type (iii). This conclusion is in agreement with the fact, established in Volume I of this book (in section 10 of Chapter 8), that for practical purposes it makes no difference to the value of a family income benefit whether the payments are made at quarterly intervals calculated from the date of death (the last payment being proportionate) or at quarterly intervals calculated from the date of the policy (the first payment being proportionate).

To summarize the results deduced in this section, we may say that the value of an annuity of type (i) may be taken as being

$$a_y - a_{xy}$$

if the annuity is curtate or

$$a_y - a_{xy} + \frac{1}{2m}(\bar{A}_y - \bar{A}_{xy}^{\ 1})$$

if it is complete, and that the value of an annuity of type (ii) or (iii) may be taken as being

$$a_y - a_{xy} - \frac{1}{2m}\bar{A}_{xy}^1$$

if it is curtate or

$$a_y - a_{xy} - \frac{1}{2m}(\bar{A}_{xy} - \bar{A}_y) \doteqdot (a_y - a_{xy})\left(1 - \frac{\delta}{2m}\right)$$

if it is complete.

25·4. Temporary and deferred reversionary annuities, etc.

The value of a continuous temporary reversionary annuity to (y) after (x), no payment being made after the expiration of n years, is given by

$$\int_0^n v^t(1 - {}_tp_x)\,{}_tp_y\,dt = \bar{a}_{y:\overline{n}|} - \bar{a}_{xy:\overline{n}|}. \tag{25·15}$$

The value of a continuous deferred reversionary annuity to (y) after (x), no payment to be made until after the expiration of n years, is

$$\int_n^\infty v^t(1 - {}_tp_x)\,{}_tp_y\,dt = {}_{n|}\bar{a}_y - {}_{n|}\bar{a}_{xy}. \tag{25·16}$$

Sometimes an annuity is required under which no payment will be made unless (x) dies within n years, leaving (y) surviving, but the annuity is to continue during the remainder of the lifetime of (y). The value of such an annuity, payable continuously, would be

$$\int_0^n v^t\,{}_tp_{xy}\,\mu_{x+t}\,\bar{a}_{y+t}\,dt \tag{25·17}$$

$$= \int_0^\infty v^t\,{}_tp_{xy}\,\mu_{x+t}\,\bar{a}_{y+t}\,dt - v^n\,{}_np_{xy}\int_0^\infty v^t\,{}_tp_{x+n:\,y+n}\,\mu_{x+n+t}\,\bar{a}_{y+n+t}\,dt$$

$$= \bar{a}_y - \bar{a}_{xy} - v^n\,{}_np_{xy}\,(\bar{a}_{y+n} - \bar{a}_{x+n:\,y+n}). \tag{25·18}$$

Alternatively, the value can be written

$$\int_0^n v^t(1 - {}_tp_x)\,{}_tp_y\,dt + \int_n^\infty v^t(1 - {}_np_x)\,{}_tp_y\,dt, \tag{25·19}$$

which reduces to the same expression.

Temporary and deferred reversionary annuities, and reversionary annuities which are payable only if (x) dies within a certain number of years, do not often occur in practice. When they do occur, they are usually payable more frequently than yearly and sufficiently accurate results will probably be obtained by regarding them as payable continuously. Should a more accurate expression be required it can be developed on the principles already set out in the chapter. It will be sufficient here to give one illustration.

If an expression is required for the value of a yearly curtate annuity of 1 to (y) after (x), payable only if (x) dies during the first n years but continuing during the whole of the remainder of (y)'s lifetime, and if this annuity conforms to type (iii) (i.e. if the first payment of 1 is to fall due one year after (x)'s death) an expression may be written down on the lines of formula (25·17) thus:

$$\int_0^n v^t {}_t p_{xy} \mu_{x+t} (\bar{a}_{y+t} - \tfrac{1}{2})\, dt.$$

This reduces to the expression given in formula (25·18) with $\frac{1}{2} \cdot {}_n\bar{A}^{1}_{xy}$ deducted from it. If the annuity conforms to type (i) (i.e. if the first payment of 1 is to fall due at the end of the year, calculated from the present time, in which (x) dies) an expression may be obtained on the lines of formula (25·19) but with integration replaced by summation, thus:

$$\sum_{t=1}^{n} v^t(1 - {}_t p_x)\,{}_t p_y + \sum_{t=n+1}^{\infty} v^t(1 - {}_n p_x)\,{}_t p_y.$$

This reduces to

$$a_y - a_{xy} - v^n\,{}_n p_{xy}(a_{y+n} - a_{x+n:y+n})$$

which is, for practical purposes, the same as formula (25·18).

25·5. Reversionary annuities involving three or more lives

Reversionary annuities involving three or more lives seldom arise in practice. When the value of such an annuity is required the development of a suitable expression presents no especial difficulty.

Two alternative methods were used in section 1 to obtain an expression for the simple reversionary annuity $\bar{a}_{x|y}$, namely

$$(a)\quad \bar{a}_{x|y}=\int_0^\infty v^t\,{}_tp_{xy}\,\mu_{x+t}\,\bar{a}_{y+t}\,dt$$

and

$$(b)\quad \bar{a}_{x|y}=\int_0^\infty v^t(1-{}_tp_x)\,{}_tp_y\,dt=\bar{a}_y-\bar{a}_{xy}\doteqdot a_y-a_{xy}.$$

When we consider the general reversionary annuity commencing on the failure of a given status (u) and continuing thereafter during the remainder of another given status (v), the solution obtained by method (b),

$$\bar{a}_{u|v}=\bar{a}_v-\bar{a}_{uv}$$

is always applicable and, if no terms certain are involved, the approximate relationship

$$\bar{a}_{u|v}=\bar{a}_v-\bar{a}_{uv}\doteqdot a_v-a_{uv}$$

applies. For example, an annuity payable continuously to (z) after the failure of the joint lifetime of (x) and (y) can be evaluated by the formula

$$\bar{a}_{xy|z}=\bar{a}_z-\bar{a}_{xyz}\doteqdot a_z-a_{xyz}. \tag{25·20}$$

If the annuity is payable to (z) after the death of the survivor of (x) and (y) we have

$$\bar{a}_{\overline{xy}|z}=\bar{a}_z-\bar{a}_{\overline{xy}:z}.$$

$$\therefore\quad \left.\begin{aligned}\bar{a}_{\overline{xy}|z}&=\bar{a}_z-\bar{a}_{xz}-\bar{a}_{yz}+\bar{a}_{xyz}\\ &\doteqdot a_z-a_{xz}-a_{yz}+a_{xyz}.\end{aligned}\right\} \tag{25·21}$$

Method (a) can be employed also but the expressions, when last survivor statuses are involved, are complicated and do not lend themselves to evaluation.

When the payment of a reversionary annuity is contingent upon the order in which the lives making up the status (u) die it is called a contingent survivorship annuity. For example, an annuity payable continuously to (z) after the failure of the joint lifetime of (x) and (y), provided that (x) dies first, is a contingent survivorship annuity and is represented by the symbol $\bar{a}^{1}_{xy|z}$. Method (a) must be used to evaluate such an annuity and we have

$$\bar{a}^{1}_{xy|z}=\int_0^\infty v^t\,{}_tp_{xyz}\,\mu_{x+t}\,\bar{a}_{z+t}\,dt. \tag{25·22}$$

If the annuity is payable to (z) after the death of the survivor of (x) and (y), provided that (y) dies second, its value is

$$\int_0^\infty v^t(1-{}_tp_x)\,{}_tp_{yz}\,\mu_{y+t}\;\bar{a}_{z+t}\,dt=\bar{a}_{y|z}-\bar{a}_{x\overset{1}{y}|z}.$$

Since, by means of method (a), $\bar{a}_{xx|y}$ can be written

$$\int_0^\infty v^t\,{}_tp_{xxy}(\mu_{x+t}+\mu_{x+t})\,\bar{a}_{y+t}\,dt,$$

we have $$\bar{a}^{1}_{xx|y}=\tfrac{1}{2}\bar{a}_{\dot{x}x|y}=\tfrac{1}{2}(\bar{a}_y-\bar{a}_{xxy}).$$

These last two results can be obtained also by general reasoning.

25·6. Annual premiums

Reversionary annuities are often issued subject to annual premiums. The net annual premium for a reversionary annuity to (y) after the death of (x) is

$$\frac{a_{x|y}}{\ddot{a}_{xy}}.$$

Net annual premiums for reversionary annuities involving three or more lives can be obtained on the lines set out in Chapter 24 for contingent assurances.

25·7. Selection

All the formulae developed in this chapter can easily be extended to select tables by merely inserting the letters x, y and z in square brackets throughout.

25·8. Office premiums

Office single and annual premiums for reversionary annuities are usually obtained by applying to the select net premiums loadings similar to those employed for temporary and contingent assurances.

25·9. Evaluation of reversionary annuities

The formulae developed in this chapter (except those involving contingent assurances and contingent survivorship annuities) can be evaluated without difficulty if single-life and joint-life annuity values are available. The evaluation of reversionary annuities where annuity values are not available and the evaluation of contingent assurances and contingent survivorship annuities will be discussed in the next two chapters.

EXAMPLES

Example 25·1

Obtain an expression for an annuity of 1 commencing on the death of (x) and payable continuously so long as either (y) or (z) survives.

Solution

If we use method (a) of section 1 we have three possibilities to consider:

(1) (y) and (z) are both alive when the annuity is entered upon,

(2) (y) is alive and (z) is dead when the annuity is entered upon,

(3) (y) is dead and (z) is alive when the annuity is entered upon.

The integral which represents the value of the annuity under each of these circumstances can be written down and we have

$$\begin{aligned}\bar{a}_{x|\overline{yz}} &= \int_0^\infty v^t {}_tp_{xyz}\,\mu_{x+t}\,\bar{a}_{\overline{y+t:z+t}}\,dt\\ &\quad+\int_0^\infty v^t(1-{}_tp_z)\,{}_tp_{xy}\,\mu_{x+t}\,\bar{a}_{y+t}\,dt\\ &\quad+\int_0^\infty v^t(1-{}_tp_y)\,{}_tp_{xz}\,\mu_{x+t}\,\bar{a}_{z+t}\,dt\\ &=(\bar{a}^{\,1}_{xz|y}+\bar{a}^{\,1}_{xy|z}-\bar{a}_{yz}+\bar{a}_{xyz})+(\bar{a}_y-\bar{a}_{xy}-\bar{a}^{\,1}_{xz|y})+(\bar{a}_z-\bar{a}_{xz}-\bar{a}^{\,1}_{xy|z})\\ &=\bar{a}_y+\bar{a}_z-\bar{a}_{yz}-\bar{a}_{xy}-\bar{a}_{xz}+\bar{a}_{xyz}.\end{aligned}$$

The same result can be obtained more easily by method (b) of section 1 thus:

$$\begin{aligned}\bar{a}_{x|\overline{yz}}&=\bar{a}_{\overline{yz}}-\bar{a}_{x:\overline{yz}}\\ &=\bar{a}_y+\bar{a}_z-\bar{a}_{yz}-\bar{a}_{xy}-\bar{a}_{xz}+\bar{a}_{xyz}.\end{aligned}$$

Example 25·2

A continuous reversionary annuity of 1 payable to (y) after the death of (x) is effected at annual premiums. Give an expression for the net premium reserve at the end of t years on the assumption that (x) is still alive.

Solution

The annual premium for the policy is

$$\frac{\bar{a}_{x|y}}{\ddot{a}_{xy}}.$$

At the end of t years the value of the annuity is

$$\bar{a}_{x+t|y+t}$$

and the value of the future premiums is

$$\frac{\bar{a}_{x|y}}{\ddot{a}_{xy}}\ddot{a}_{x+t:y+t}.$$

Hence the net premium reserve is

$$\bar{a}_{x+t|y+t}-\frac{\bar{a}_{x|y}}{\ddot{a}_{xy}}\ddot{a}_{x+t:y+t}.$$

If we use the approximation

$$\bar{a}_{x|y} \doteq a_y - a_{xy} = \ddot{a}_y - \ddot{a}_{xy},$$

this expression becomes

$$\ddot{a}_{y+t}-\ddot{a}_{x+t:y+t}-(\ddot{a}_y-\ddot{a}_{xy})\frac{\ddot{a}_{x+t:y+t}}{\ddot{a}_{xy}}=\ddot{a}_{y+t}-\ddot{a}_y\frac{\ddot{a}_{x+t:y+t}}{\ddot{a}_{xy}}.$$

Example 25·3

An annuity of 1 is payable until the last death of three lives now aged x, y and z. The whole annuity is payable to (x) during his life, and on his death (y) and (z) are to receive equal shares while both are living, the survivor to take the whole, subject, however, to the proviso that at the death of (y) his share shall be payable to his two sons u and v (in equal shares while both are living and the whole to the survivor).

Express in terms of immediate annuities the value of the share of each of the five lives and prove your results.

Solution

It will simplify matters if we first find an expression for a reversionary annuity of A payable to (y) so long as he is alive jointly with (z) after the death of (x), increasing to B payable to (y) after the death of the survivor of (x) and (z). This can be expressed in terms of reversionary annuities in three ways:

$$Aa_{x|yz}+Ba_{\overline{xz}|y}. \tag{1}$$

$$Aa_{x|y}+(B-A)\,a_{\overline{xz}|y}. \tag{2}$$

$$Ba_{x|y}-(B-A)\,a_{x|yz}. \tag{3}$$

Of these three expressions, (1) is the easiest to write down, but (3) is the easiest to develop in terms of immediate annuities thus:

$$B(a_y-a_{xy})-(B-A)\,(a_{yz}-a_{xyz}). \tag{4}$$

(x)'s share $\qquad a_x.$

(y)'s share $\qquad \tfrac{1}{2}a_{x|yz}+a_{\overline{xz}|y}.$

By reference to (1) and (4) we see at once that this is equal to

$$(a_y-a_{xy})-\tfrac{1}{2}(a_{yz}-a_{xyz}).$$

(z)'s share $\qquad \tfrac{1}{2}a_{x|\overline{yuv}:z}+a_{\overline{xyuv}|z}.$

By reference to (1) and (4), writing $\overline{yuv}$ for z and z for y, we have

$$(a_z-a_{xz})-\tfrac{1}{2}(a_{\overline{yuv}:z}-a_{x:\,\overline{yuv}:z}).$$

(u)'s share $\qquad \tfrac{1}{4}a_{\overline{xy}|uvz}+\tfrac{1}{2}a_{\overline{xyv}|uz}.$

By reference to (1) and (4) again, we have

$$\tfrac{1}{2}(a_{uz}-a_{\overline{xy}:uz})-\tfrac{1}{4}(a_{uvz}-a_{\overline{xy}:uvz}).$$

(v)'s share. Similarly, this is

$$\tfrac{1}{2}(a_{vz}-a_{\overline{xy}:vz})-\tfrac{1}{4}(a_{uvz}-a_{\overline{xy}:uvz}).$$

By developing the last survivor statuses in the usual way and cancelling out we get as the total

$$a_x+a_y+a_z-a_{xy}-a_{xz}-a_{yz}+a_{xyz}=a_{\overline{xyz}}.$$

CHAPTER 26

EVALUATION OF FUNCTIONS OF TWO OR MORE LIVES—COMMUTATION FUNCTIONS AND APPROXIMATE INTEGRATION

26·1. Introduction

When a mortality table is to be employed at all widely it is worth while to tabulate extensive sets of single-life commutation functions, annuity values, etc., by means of which any functions which are not tabulated may be easily evaluated. Even if the table is to have only a limited use, the labour involved in constructing single-life commutation columns is usually amply justified by the usefulness of these columns when the values of annuities and assurances are required.

When we come to consider the evaluation of joint-life and contingent assurance functions we find that the position is very different. Whereas, for example, a complete table of numerical values of D_x, based on ultimate mortality and one rate of interest, can normally be accommodated quite easily in two columns of a printed book, the corresponding table of D_{xy}, covering all possible combinations of x and y, would occupy many pages. The tabulation of numerical values of a function such as D_{xyz}, for all possible combinations of x, y and z, would obviously be out of the question.

In this chapter it is proposed to consider the suitability of various methods of partial tabulation. We shall also discuss the use of approximate integration for the purpose of obtaining individual values of annuities and assurances when suitable functions are not tabulated at all.

26·2. Commutation functions for the joint-life mortality table

Before we proceed to problems of tabulation, it is necessary to consider the form which is taken by the various joint-life commutation functions.

The first function to be considered is D_{xy}. According to the International Actuarial Notation, D_{xy} is defined by the relationship

$$D_{xy}=v^{\frac{1}{2}(x+y)}\,l_{xy}. \tag{26·1}$$

In practical work, however, it is often found convenient to depart from the International Notation and to write

$$\left.\begin{aligned} D_{xy}&=v^{x}\,l_{xy}\\ \text{or}\qquad D_{xy}&=D_{x}\,l_{y}.\end{aligned}\right\} \tag{26·2}$$

Ambiguity can be avoided, as suggested in Chapter 21, by defining (x) as the life whose age is mentioned first in the suffix of the joint-life symbol. When one life is an 'assured' life and the other is an 'annuitant' life, it is customary to allot the letter x to the 'assured' life and y to the 'annuitant' life; when one life is a male and the other is a female, it is customary to use x for the male life and y for the female.

We have seen in Chapter 22 that

$${}_{t}E_{xy}=v^{t}\,{}_{t}p_{xy}.$$

It follows that, whether D_{xy} is defined by formula (26·1) or by formula (26·2),

$${}_{t}E_{xy}=\frac{D_{x+t:\,y+t}}{D_{xy}}.$$

The value of a joint-life annuity of 1 to (x) and (y) is

$$a_{xy}=\frac{N_{x+1:\,y+1}}{D_{xy}}, \tag{26·3}$$

where

$$N_{xy}=\sum_{t=0}^{\infty}D_{x+t:\,y+t}. \tag{26·4}$$

Similarly

$$(Ia)_{xy}=\frac{S_{x+1:\,y+1}}{D_{xy}},$$

where

$$S_{xy}=\sum_{t=0}^{\infty}N_{x+t:\,y+t}. \tag{26·5}$$

S_{xy} is rarely evaluated.

For assurances we require C_{xy}. This is defined in the International Actuarial Notation by the relationship

$$C_{xy}=v^{\frac{1}{2}(x+y)+1}\,d_{xy}. \tag{26·6}$$

If, however, D_{xy} has been obtained by formula (26·2), it is necessary to obtain C_{xy} by the formula

$$C_{xy} = v^{x+1} d_{xy}. \tag{26·7}$$

It should be noted that if C_{xy} is not being evaluated and if, as will usually be the case, numerical values of D_x are available, formula (26·2) has the great advantage that, in its second form, it requires only one multiplication whereas formula (26·1) requires two (the multiplication of l_x by l_y to obtain l_{xy} and the multiplication of l_{xy} by $v^{\frac{1}{2}(x+y)}$). If, however, numerical values of C_{xy} are required, it will be necessary to evaluate l_{xy} in order to obtain d_{xy} $(=l_{xy} - l_{x+1:y+1})$ and this advantage of formula (26·2) will disappear. Even in these circumstances formula (26·2) may still be preferred, partly because it involves only integral powers of v and partly because, if a table of $D_{x:y+r}$ (x and y constant, r varying) is required, v^x or D_x can be set up on the calculating machine and treated as a constant multiplier. Similarly, formula (26·7) may be preferred to formula (26·6), partly because it involves only integral powers of v and partly because, if a table of $C_{x:y+r}$ is required, v^{x+1} can be set up on the calculating machine as a constant multiplier.

The other commutation functions for assurances are

$$M_{xy} = \sum_{t=0}^{\infty} C_{x+t:y+t} \tag{26·8}$$

and

$$R_{xy} = \sum_{t=0}^{\infty} M_{x+t:y+t}. \tag{26·9}$$

The value of a joint whole-life assurance of 1 payable at the end of the year of the first death of (x) and (y) is

$$A_{xy} = \frac{M_{xy}}{D_{xy}}.$$

The value of an increasing assurance, under which the sum assured during the tth year is t, is

$$(IA)_{xy} = \frac{R_{xy}}{D_{xy}}.$$

The annual premiums for these assurances are

$$P_{xy} = \frac{M_{xy}}{N_{xy}}$$

and $$P(IA)_{xy}=\frac{R_{xy}}{N_{xy}},$$
respectively.

When three or more lives are involved we may write

$$\left.\begin{aligned} D_{xyz}&=v^x\,l_{xyz}\\ \text{or}\qquad D_{xyz}&=D_x\,l_y\,l_z,\\ N_{xyz}&=\sum_{t=0}^{\infty}D_{x+t\,:\,y+t\,:\,z+t}\end{aligned}\right\}\qquad(26\cdot10)$$

and so on. However, functions for three or more lives are not normally evaluated unless the ages of all the lives are equal. Thus we have

$$\left.\begin{aligned} D_{xxx}&=v^x\,l_{xxx}\\ \text{or}\qquad D_{xxx}&=D_x\,l_x\,l_x,\\ N_{xxx}&=\sum_{t=0}^{\infty}D_{x+t:\,x+t:\,x+t}\end{aligned}\right\}\qquad(26\cdot11)$$

and so on.

When select tables are used, the same principles apply. For example, formula (26·1) becomes

$$D_{[xy]}=v^{\frac{1}{2}(x+y)}\,l_{[xy]},$$

formula (26·2) becomes

$$D_{[xy]}=v^x\,l_{[xy]}$$

or $$D_{[xy]}=D_{[x]}\,l_{[y]},$$

and so on.

Expressions in commutation form for other annuities and assurances involving two or more lives can be written down without difficulty; it is hardly necessary to give further examples here.

26·3. Tabulation of joint-life functions

If a published mortality table is to be used for calculations in respect of joint-life assurances, it is customary to construct fairly extensive tables of joint-life functions for equal ages. For example, the A 1924–29 Table contains values of D_{xx}, N_{xx}, M_{xx}, R_{xx}, a_{xx}, A_{xx} and P_{xx} together with the corresponding select functions. If the mortality table is to be used mainly in connexion with annuities rather than assurances, tabulation of M_{xx}, R_{xx}, A_{xx} and P_{xx}, with their corresponding select functions, will hardly be necessary, especially as M, A and P can be obtained, when required, by the premium conversion relationships.

The question of the extent to which it is necessary to tabulate functions for two lives of different ages is an interesting one. We shall see in the next chapter that if the mortality table follows Gompertz's or Makeham's 'law', numerical values of functions in respect of two lives of different ages can be obtained quite easily from those in respect of two lives of equal age. In such circumstances, tabulation of functions in respect of two lives of different ages can be completely avoided without detriment to the usefulness of the tables. If the mortality table does not follow either of these laws, some tabulation in respect of two lives of different ages will usually be necessary. It may be impracticable to tabulate even a limited number of functions, such as D_{xy}, N_{xy} and a_{xy}, for all possible combinations of x and y.

It would, of course, be possible to tabulate the functions for every fifth age, and to use an interpolation formula to obtain values of the functions for other ages. The objection to this is that if the tables are to be in frequent use it must be possible to obtain any required figures from them with the expenditure of very little time and labour. Some compromise must therefore be found between the two extremes of tabulating the functions for all possible combinations of ages (which permits of the extraction of any required figures without further calculation) and tabulating them for such limited combinations of ages that the process of using the tables involves elaborate calculations.

A device which was adopted in the $a(f)$ and $a(m)$ tables, published in 1924, was to tabulate $a_{[xy]}$, a_{xy}, $a_{[\overline{xy}]}$ and $a_{\overline{xy}}$ for every third age. This method was followed, in respect of $a_{[xy]}$ and a_{xy}, in the A 1924–29 tables. The intention is that any required value which is not tabulated should be obtained by first difference interpolation between two tabulated values. The principle may be illustrated by the diagram on page 117, in which x, $x+3$, y and $y+3$ are assumed to be ages in respect of which joint-life annuity values (represented by the letter a) are tabulated. Interpolation is made between the two corner points which lie at the ends of the line passing through the point for which the value is required. For example

$$a_{x:y+1}=\tfrac{2}{3}a_{xy}+\tfrac{1}{3}a_{x:y+3},$$

$$a_{x+1:y+2}=\tfrac{2}{3}a_{x:y+3}+\tfrac{1}{3}a_{x+3:y}$$

and so on.

It will be appreciated that although this method of interpolation gives sufficiently accurate results for most practical purposes it leads to irregularities when a series of values is required. Moreover, it sometimes produces anomalous results. A further objection to the method is that the process of interpolation does not lend itself

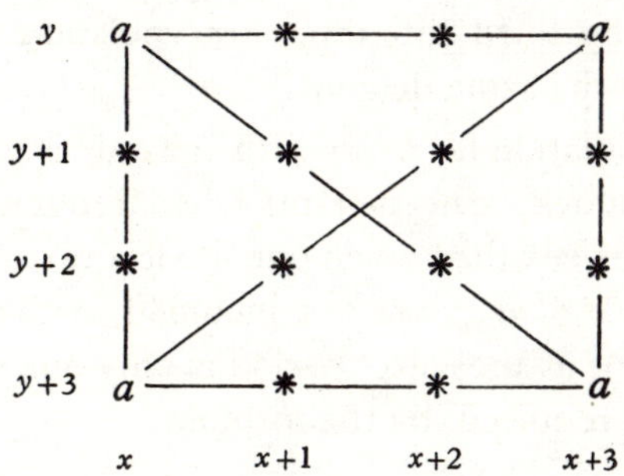

to the use of a calculating machine and has to be performed mentally; a consequence of this is that if a large number of values is required some workers find the method rather tedious and conducive to error. However, it has the great advantage that the table of annuity values occupies little more than one-ninth of the space that would be required if values for all ages were tabulated. This is an important consideration especially when (as in the case of the *a*(*f*) and *a*(*m*) tables) three sets of values of $a_{[xy]}$, a_{xy}, $a_{[\overline{xy}]}$ and $a_{\overline{xy}}$ at various rates of interest are required, one set being applicable when (x) and (y) are both females, one when they are both males and one when (x) is a male and (y) is a female.

In the *a*(55) tables, published in 1953, values of a_{xy}, $a_{[\overline{xy}]}$ and $a_{\overline{xy}}$ (but not $a_{[xy]}$) are tabulated for even ages when (x) is a male and (y) is a female; values for other ages can be obtained by first-difference interpolation. When (x) and (y) are both males, or both females, the functions are tabulated for every fifth age; values for other ages may conveniently be obtained by interpolation as follows:

(i) if the required value does not lie in a row, column or diagonal of the table, by a formula which is equivalent to interpolating first horizontally and then vertically (or first vertically and then horizontally) between the four tabulated values at the corners of the square in which the required value lies,

(ii) if the required value lies in a row or column of the table, by first difference interpolation between two tabulated values, or by third difference interpolation between four tabulated values, in the same row or column,

(iii) if the required value lies in a diagonal of the table, either by method (i) or by third difference interpolation between four tabulated values in the same diagonal.

In these tables, tabulation for every fifth age only is justified by the fact that annuity values are needed much less frequently when both lives are of the same sex than when one life is a male and the other is a female. Values of $a_{[xy]}$ are not included in the $a(55)$ tables (except when $y=x$); as the select period is only one year, they can be obtained, when required, by the formula

$$a_{[xy]} = a_{xy}\frac{p_{[xy]}}{p_{xy}}.$$

It is not usual to tabulate values of temporary joint-life annuities, even for two lives of equal age. Tabulation for two lives of different ages would be impracticable. Values may be obtained by the formula

$$a_{xy:\overline{n}|} = a_{xy} - \frac{D_{x+n:\,y+n}}{D_{xy}}\,a_{x+n:\,y+n}. \tag{26·12}$$

Single and annual premiums for endowment assurances can be obtained from the temporary annuities by the premium conversion relationships. Single premiums for temporary assurances can be obtained by the formula

$$A^{\;1}_{\overline{xy}:\overline{n}|} = A_{xy:\overline{n}|} - \frac{D_{x+n:\,y+n}}{D_{xy}}, \tag{26·13}$$

or, if n is small, the approximation

$$A^{\;1}_{\overline{xy}:\overline{n}|} \doteq A^{1}_{x:\overline{n}|} + A^{1}_{y:\overline{n}|} \tag{26·14}$$

can be used. Select functions may be obtained similarly.

26·4. Commutation functions for contingent assurances

Commutation functions suitable for the evaluation of contingent assurances are C^1_{xy} and M^1_{xy}.

According to the International Actuarial Notation,

$$C^1_{xy} \doteq v^{\frac{1}{2}(x+y)+1} d_x l_{y+\frac{1}{2}}, \tag{26·15}$$

but if D_{xy} has been obtained by formula (26·2) it is necessary to write

$$\left.\begin{aligned} &C^1_{xy} \doteq v^{x+1} d_x l_{y+\frac{1}{2}} \\ \text{or}\quad &C^1_{xy} \doteq C_x l_{y+\frac{1}{2}}. \end{aligned}\right\} \tag{26·16}$$

Formula (26·16), in its second form, is very convenient if, as will usually be the case, C_x has already been evaluated. $l_{y+\frac{1}{2}}$ may be taken as being equal to $\frac{1}{2}(l_y + l_{y+1})$. It is usual to employ 'assured lives' mortality for the assured life (x), and 'annuitants' mortality for the counter life (y).

The value of an assurance of 1 payable at the end of the year of death of (x), if he predeceases (y), is

$$A^1_{xy} = \frac{M^1_{xy}}{D_{xy}},$$

where

$$M^1_{xy} = \sum_{t=0}^{\infty} C^{\;1}_{x+t:y+t}. \tag{26·17}$$

The annual premium for this assurance is

$$P(A^1_{xy}) = \frac{M^1_{xy}}{N_{xy}}.$$

It was pointed out in Chapter 24 that there is little advantage, in the case of contingent assurances, in assuming that the sum assured is payable at the end of the year of death. If it is desired to calculate the values of functions on the assumption that the sum assured is payable immediately on the death of (x), and if D_{xy} has been obtained by formula (26·2), a convenient procedure is to obtain $\bar{C}^1_{xy}$ by the formula

$$\left.\begin{aligned} &\bar{C}^1_{xy} \doteq v^{x+\frac{1}{2}} d_x l_{y+\frac{1}{2}} \\ \text{or}\quad &\bar{C}^1_{xy} \doteq \bar{C}_x l_{y+\frac{1}{2}}, \end{aligned}\right\} \tag{26·18}$$

and hence

$$\bar{M}^1_{xy} = \sum_{t=0}^{\infty} \bar{C}^{\;1}_{x+t:y+t}. \tag{26·19}$$

It is hardly necessary to set out here the expressions for $\bar{A}^1_{xy}$ and $P(\bar{A}^1_{xy})$ in commutation form.

Contingent assurance functions are often required on a select basis; the adaptation of the foregoing formulae to select tables presents no difficulty.

Commutation functions for contingent assurances involving three or more lives are not normally required, since such assurances are more conveniently evaluated by approximate methods.

26·5. Tabulation of contingent assurance functions

It is many years since tables of contingent assurance functions were published in Great Britain (except that tables of $\ddot{a}_{xy}$ and $\bar{A}^1_{xy}$ on up-to-date mortality and at one rate of interest only, for triennial ages, together with tables for obtaining the corresponding select functions approximately, have been published by C. M. O'Brien in *J.I.A.* **80** [1954], 402). Should it be desired to construct such tables on up-to-date mortality, it would be necessary to decide, first, what functions to include in the tables and, secondly, for what combinations of ages these functions should be tabulated. The functions chosen might well be $a_{[xy]}$, a_{xy}, $\bar{A}^1_{[x]:[y]}$, $\bar{A}^1_{xy}$ and $P(\bar{A}^1_{[x]:[y]})$. They could be tabulated:

(i) for all integral ages,
(ii) for even ages only,
(iii) for every third age,
or (iv) for every fifth age.

As contingent assurance functions are not often required, tabulation for every fifth age would probably be considered sufficient.

26·6. Approximate integration—direct method

When it is desired to evaluate an annuity or assurance which is not tabulated and for which commutation functions are not available, recourse may be had to approximate integration. This method is suitable when an individual value is required but it is too laborious, generally speaking, for the calculation of a series of values.

The student will find a development of the more common approximate integration formulae in Chapter 9 of *Mathematics for Actuarial*

Students, Part II, by H. Freeman, but it will be convenient to reproduce some of the formulae here.

Hardy's '39*a*'

$$\int_0^\infty u_t dt = n(\cdot 28u_0 + 1\cdot 62u_n + 2\cdot 2u_{3n} + 1\cdot 62u_{5n} + \cdot 56u_{6n} + 1\cdot 62u_{7n}). \tag{26·20}$$

This formula is known as '39*a*' because it was so numbered in *The Institute of Actuaries' Text Book*, Part II, *Life Contingencies* [1887] by G. King. It was obtained from the following formula, which had been suggested by G. F. Hardy in *J.I.A.* **24** [1883], 95:

$$\int_0^{6n} u_t dt = n(\cdot 28u_0 + 1\cdot 62u_n + 2\cdot 2u_{3n} + 1\cdot 62u_{5n} + \cdot 28u_{6n}).$$

If this formula, which is exact if u_t is a polynomial of the 5th degree, is applied to the range 0 to $6n$, and also to the range $6n$ to $12n$, and if u_{9n}, u_{11n}, u_{12n} and all higher values are then made equal to zero, formula 39*a* results.

Weddle

$$\int_0^{6n} u_t dt = \frac{3n}{10}(u_0 + 5u_n + u_{2n} + 6u_{3n} + u_{4n} + 5u_{5n} + u_{6n}). \tag{26·21}$$

This formula is exact if u_t is a polynomial of the 5th degree throughout the whole range. It is very nearly correct for a polynomial of the 7th degree.

Simpson

$$\int_0^{2n} u_t dt = \frac{n}{3}(u_0 + 4u_n + u_{2n}). \tag{26·22}$$

This formula is exact if u_t is a polynomial of the 3rd degree.

Repeated Simpson

$$\int_0^{2nr} u_t dt = \frac{n}{3}(u_0 + 4u_n + 2u_{2n} + 4u_{3n} + \dots + u_{2nr}). \tag{26·23}$$

This formula is exact if u_t is a polynomial of the 3rd degree for the range 0 to $2n$, another polynomial of the 3rd degree for the range $2n$ to $4n$ and so on.

The Three-eighths rule

$$\int_0^{3n} u_t\,dt = \frac{3n}{8}(u_0 + 3u_n + 3u_{2n} + u_{3n}). \qquad (26\cdot24)$$

This is exact if u_t is a polynomial of the 3rd degree.

Hardy's '39*a*' has always been popular with actuaries because it has been found, by experience, to give good results for most 'whole-life' functions. It is not suitable for temporary annuities or temporary assurances. If the table has a limiting age ω, it is usual to choose n as the integer such that $7n$ is as nearly as possible equal to the value of t at which u_t is first equal to zero. When the function to be evaluated is a joint-life annuity or a simple contingent assurance payable on the first death, this means that n must be so chosen that when $7n$ is added to the age of the oldest life the resulting age lies just within or just beyond the limit of the table. When the function is a last survivor annuity, so that the integrand takes the form $v^t\{1-(1-{}_tp_x)(1-{}_tp_y)\ldots\}$, $7n$ must be added to the age of the *youngest* life, because the integrand does not become zero until the age of the youngest life is equal to the limiting age of the table. If the table has no limiting age ω, some arbitrary assumption must be made for the purpose of determining n. A convenient assumption is that $\omega-1$ is the first integral age at which $q_x > \frac{1}{2}$.

'Weddle' and 'Repeated Simpson' are suitable either for whole-life functions or for temporary functions with a fairly long term. 'Simpson' and the 'Three-eighths rule' are suitable when the term is short.

As it is clearly an advantage that the values of u_n that are used in the calculations should relate to integral ages, the choice of a formula for a temporary function will be limited by the extent to which the term of the function can be factorized. For example, it is not convenient to use 'Weddle' unless the term is divisible by 6, 'Simpson' unless it is divisible by 2, 'Repeated Simpson' unless it is divisible by 4, 6 or 10, or the 'Three-eighths rule' unless it is divisible by 3.

It will be seen that if the term of a temporary function is 25, 35, 55, ... years (i.e. a multiple of 5 but not a multiple of 2 or 3) none of the formulae given above is suitable. As 25 years is by no means an unusual term for actuarial functions, this is a little unfortunate. M. T. L. Bizley drew attention to this fact in *J.S.S.* **6** [1946], 90 and suggested the following formula which is very nearly accurate provided that sixth differences are zero:

$$\int_0^{5n} u_t dt = n(\cdot 33u_0 + 1\cdot 3u_n + \cdot 87u_{2n} + \cdot 87u_{3n} + 1\cdot 3u_{4n} + \cdot 33u_{5n}). \tag{26·25}$$

In selecting a suitable formula for evaluating a life-table function, it is important to bear in mind that a formula which is theoretically correct to a high order of differences will not necessarily give a more accurate result than one which is correct to only a low order. The reason for this is that while most life-table functions approximate to polynomials of low degree over short sections of the mortality table, very few of them approximate to polynomials of high degree over longer ranges. It follows that a formula such as 'Repeated Simpson', which is true only to third differences, often gives better results than other formulae which are theoretically correct to sixth and seventh differences.

When the function to be integrated is discontinuous, as when a select table is employed, it is necessary, strictly speaking, to evaluate separately the two portions on either side of the point of discontinuity. In practice, however, this refinement can often be ignored, especially if it can be arranged that one of the values used in the approximate integration occurs at or near the point of discontinuity.

26·7. An example of approximate integration—direct method

As an example of approximate integration, we may take the evaluation, by Hardy's formula '39*a*', of a contingent assurance payable on the death of (30) if he dies before (45). In practice two different mortality tables would probably be used but, for the purpose of illustrating the method, all the functions will be taken from the A 1924–29 Ultimate table. The rate of interest used will be $2\frac{1}{2}$ per cent.

The calculation will be made with the aid of logarithms in order to show how an approximate integration may conveniently be performed in the absence of a calculating machine. In modern usage a calculating machine would normally be employed.

$$\bar{A}^{1}_{30:45} = \frac{1}{D_{30} l_{45}} \int_0^\infty D_{30+t} l_{45+t} \mu_{30+t} \, dt.$$

$$q_{103} = \cdot 49211,$$

$$q_{104} = \cdot 51420.$$

$$\therefore \quad \omega = 105 \quad \text{say},$$

$$n = (105 - 45) \div 7 = 8\cdot 57 = 9 \quad \text{say}.$$

The working may be set out as follows:

t	0	9	27	45	54
$\log D_{30+t}$	—	6·55190	6·30353	5·80227	5·20005
$\log l_{45+t}$	—	6·93045	6·70279	5·49527	3·97244
log coefft	$\bar{1}\cdot 44716$	0·20952	0·34242	0·20952	$\bar{1}\cdot 74819$
$-\log D_{30} l_{45}$	—	$\overline{14}\cdot 38165$	$\overline{14}\cdot 38165$	$\overline{14}\cdot 38165$	$\overline{14}\cdot 38165$
Sub-total	$\bar{1}\cdot 44716$	0·07352	$\bar{1}\cdot 73039$	$\bar{3}\cdot 88871$	$\bar{5}\cdot 30233$
$\log \mu_{30+t}$	$\bar{3}\cdot 37840$	$\bar{3}\cdot 54777$	$\bar{2}\cdot 14208$	$\bar{2}\cdot 92829$	$\bar{1}\cdot 26503$
Total	$\bar{4}\cdot 82556$	$\bar{3}\cdot 62129$	$\bar{3}\cdot 87247$	$\bar{4}\cdot 81700$	$\bar{6}\cdot 56736$
Antilog	·0006692	·0041811	·0074554	·0006561	·0000037

·0006692
·0041811
·0074554
·0006561
·0000037

$$\therefore \quad \bar{A}^{1}_{30:45} = \cdot 0129655 \times 9 = \cdot 11669.$$

The true value of $\bar{A}^{1}_{30:45}$, obtained by constructing a column of $\bar{C}^{1}_{30+t:45+t}$, summing and dividing by $D_{30:45}$, is ·11554.

A sub-total has been included in the calculations above in order to show how, if $\ddot{a}_{30:45}$ is required for the calculation of the annual premium, it may be obtained in the course of the approximate integration, thus:

$$\begin{aligned}
\text{Antilog}\ \bar{1}\cdot 44716 &= \ \ \cdot 2800 \\
\text{,,}\quad 0\cdot 07352 &= 1\cdot 1845 \\
\text{,,}\quad \bar{1}\cdot 73039 &= \ \ \cdot 5375 \\
\text{,,}\quad \bar{3}\cdot 88871 &= \ \ \cdot 0077 \\
\text{,,}\quad \bar{5}\cdot 50233 &= \ \ \cdot 0000 \\
\therefore\quad \bar{a}_{30:45} &\fallingdotseq 2\cdot 0097 \times 9 = 18\cdot 087. \\
\therefore\quad \ddot{a}_{30:45} &\fallingdotseq 18\cdot 587.
\end{aligned}$$

The true value, according to the published table of joint-life annuities, is 18·537.

If the annual premium is not required, or if $a_{30:45}$ is tabulated, the sub-total may be omitted.

*26·8. Approximate integration—Product integration

The integrands with which we are concerned can be expressed as the product of two functions, $\phi(t)$ and $f(t)$, and the value of the integral can be expressed as the integral of one of the functions multiplied by a weighted sum of the values of the other function at n points. Thus

$$\int_x^y \phi(t) f(t)\, dt = \{a_1 f(t_1) + a_2 f(t_2) + \ldots + a_n f(t_n)\} \int_x^y \phi(t)\, dt.$$

The use of this type of formula for approximating to the value of actuarial functions was first suggested by W. Perks in *J.I.A.* **64** [1933], 286, for the case $a_1 = a_2 = \ldots = 1/n$. Perks showed that the values of t_r were dependent on the successive moments of $\phi(t)$. H. G. Jones in *J.I.A.* **64** [1933], 318, further developed the method and a systematic development was later made by R. E. Beard in *J.I.A.* **73** [1947], 356.

For single-life whole-life functions the formulae can be conveniently applied by taking $\phi(t)$ as $(\mu l)_{x+t}$ and interpolating in v^t for the values of $f(t)$. For joint-life whole-life functions $\phi(t)$ may be taken as $(\mu l)_{x+t}$ and $f(t)$ as $v^t l_{y+t}$ or alternatively $\phi(t)$ may be taken as $(\mu l)_{x+t:y+t}$ and $f(t)$ as v^t.

The moment functions of $(\mu l)_x$ (and the corresponding functions for two, three, four and five lives of equal ages) for the A 1924–29 table are given in Beard's paper together with tables of a_r and t_r for various formulae.

For temporary functions it is more convenient to take $\phi(t)$ as v^t and to interpolate in the life table for the other factors in the integrand; suitable tables for this purpose are given in Beard's paper.

For more complicated functions, such as contingent assurances, some care has to be used in the selection of a formula. The error term depends to a large extent on the behaviour of the function $f(t)$ and if this is a rapidly varying function a formula applied to the whole range in one operation will be liable to be unsatisfactory. In these circumstances it is desirable to divide the range of integration and to apply a formula using $\phi(t)=v^t$ to the successive parts of the integral. Experiments which have been made suggest also that if, for example, $f(t)$ is very small when t is greater than about 50 and becomes zero when $t=60$, better results are obtained by limiting the range of the approximate integration to 50 years than by applying the formula over the whole period of 60 years.

To illustrate the method, formula (91) of Beard's paper will be applied to the evaluation of the same contingent assurance as was used to illustrate Hardy's formula '39*a*' in the previous section. A calculating machine will be used, but the calculations could be made quite conveniently with the aid of logarithms if a machine were not available.

Formula (91), which might be classed as a general purpose formula convenient for the calculation of many actuarial functions, may be written

$$\int_0^m v^t f(t)\,dt=\left\{a_1 f(0)+a_2 f\left(\frac{m}{2}\right)+a_3 f(m)\right\}\bar{a}_{\overline{m}|}. \qquad (26{\cdot}26)$$

For the calculation of $\bar{A}^{\,1}_{30:45}$, since $f(t)$ is very small when $t>50$, it will be convenient to take $m=25$ and to apply the formula twice. Values of a_1, a_2 and a_3 at $2\frac{1}{2}$ per cent may be found by first-difference interpolation in the table in Beard's paper and the entire working is as follows:

t	l_{30+t}	l_{45+t}	μ_{30+t}	v^{25}	Coefft	Product $\times 10^{-7}$
0	9,574,248	9,098,372	·00239	—	·21990	4,578
12½	9,206,695	8,158,846	·00440	—	·66232	21,890
25	8,427,897	5,660,936	·01141	—	·11778	6,412
25	8,427,897	5,660,936	·01141	·53939	·21990	6,457
37½	6,343,212	1,639,186	·03990	·53939	·66232	14,821
50	2,365,827	57,314	·13240	·53939	·11778	114
						54,272

$$\therefore \quad \bar{A}^{1}_{30:45} \doteq 54{,}272 \times \frac{\bar{a}_{\overline{25}|}}{l_{30:45}} \times 10^7 = \cdot 11622.$$

It will be noted that formulae such as Simpson's or the Three-eighths rule may be obtained as special cases of the above general formula by putting $\phi(t) = 1$ for all values of t. (See Beard, *J.I.A.* **73** [1947], 373, section 58.)

EXAMPLES

Example 26·1

Find an expression for the net annual premium for a 5-year temporary assurance, payable at the end of the year in which the first death occurs, on three lives aged x, y and z, respectively.

Obtain a numerical result at 3 per cent interest, when $x = 50$, $y = 51$ and $z = 52$ given the following data:

$$\log l_{50} = 5\cdot 65720$$
$$\log l_{51} = 5\cdot 65415$$
$$\log l_{52} = 5\cdot 65078$$
$$\log l_{53} = 5\cdot 64704$$
$$\log l_{54} = 5\cdot 64291$$
$$\log l_{55} = 5\cdot 63832$$
$$\log l_{56} = 5\cdot 63323$$
$$\log l_{57} = 5\cdot 62759$$
$$\log(1+i) = \cdot 0128372$$

Solution

$$A^{\ \ 1}_{\overline{xyz}:\overline{5}|}=A_{xyz:\overline{5}|}-A_{xyz:\overline{5}|}^{\ \ \ \ \ 1}$$
$$=1-d\ddot{a}_{xyz:\overline{5}|}-(a_{xyz:\overline{5}|}-a_{xyz:\overline{4}|})$$
$$=v\ddot{a}_{xyz:\overline{5}|}-a_{xyz:\overline{5}|}.$$

Alternatively, by general reasoning, the single premium is the difference between a 5-year temporary annuity payable at the end of each year on which the joint status *xyz* enters, and a 5-year temporary annuity payable at the end of each year which the joint status completes.

In either case, the annual premium is

$$P^{\ \ 1}_{\overline{xyz}:\overline{5}|}=v-\frac{a_{xyz:\overline{5}|}}{\ddot{a}_{xyz:\overline{5}|}}.$$

Numerical result

t	1	2	3	4	5
$\log v^t$	$\bar{1}$·98716	$\bar{1}$·97433	$\bar{1}$·96149	$\bar{1}$·94865	$\bar{1}$·93581
$\log l_{50+t}$	5·65415	5·65078	5·64704	5·64291	5·63832
$\log l_{51+t}$	5·65078	5·64704	5·64291	5·63832	5·63323
$\log l_{52+t}$	5·64704	5·64291	5·63832	5·63323	5·62759
$-\log l_{50:51:52}$	$\overline{17}$·03787	$\overline{17}$·03787	$\overline{17}$·03787	$\overline{17}$·03787	$\overline{17}$·03787
Total	$\bar{1}$·97700	$\bar{1}$·95293	$\bar{1}$·92763	$\bar{1}$·90098	$\bar{1}$·87282
Antilog	·9484	·8973	·8465	·7961	·7461

$$a_{xyz:\overline{5}|}=\cdot9484+\cdot8973+\cdot8465+\cdot7961+\cdot7461=4\cdot2344,$$
$$\ddot{a}_{xyz:\overline{5}|}=1+\cdot9484+\cdot8973+\cdot8465+\cdot7961\qquad=4\cdot4883.$$
$$\therefore\quad P=\cdot97087-\cdot94343=\cdot02744.$$

Rough check

The average age of the three lives at the beginning of the third year (middle year of term) is 53.

$$\log l_{54}=5\cdot64291,$$
$$\log l_{53}=5\cdot64704.$$
$$\therefore\quad \log p_{53}=\bar{1}\cdot99587.$$
$$\therefore\quad p_{53}=\cdot9905\quad\text{and}\quad q_{53}=\cdot0095.$$

The answer is a little less than 3 times q_{53}, which is reasonable.

Example 26·2

On 31 December 1954, a sum of £10,000 was lent by two persons, each then aged 60, the loan to bear interest at 5 per cent per annum convertible half-yearly and to be repaid by equal half-yearly instalments including principal and interest over a period of 50 years. It was subsequently agreed that upon the death of the survivor the principal outstanding should be repayable immediately with interest, the necessary sum being obtained from another source.

On 1 January 1960, both being then alive, they offer to sell the balance of their interest, after retaining the right to all the payments falling due before the death of the survivor.

On the basis of the Hypothetical Select Mortality Table, ultimate mortality, with interest at 4 per cent per annum effective, find by means of an approximate integration the price they should receive.

Ignore income tax.

You are given that $\log_{10} 1{\cdot}04 = {\cdot}0170333$. You are also given the following values:

n	$a_{\overline{n}\vert}^{\cdot 025}$
10	8·7521
20	15·5892
30	20·9303
40	25·1028
50	28·3623
60	30·9087
70	32·8979
80	34·4518
90	35·6658
100	36·6141

Solution

Let F_t be the principal outstanding with interest at time t, time being calculated from 1 January 1960. Then the price

$$=2\int_0^\infty v^t F_t(1-{}_tp_{65})\,{}_tp_{65}\,\mu_{65+t}\,dt.$$

Now $$10{,}000(a_{\overline{100}\vert}^{\cdot 025})^{-1}=273{\cdot}12.$$

During each half-year F_t is increasing continuously, but at the moment when the $2t$th payment falls due it changes from $273{\cdot}12(a_{\overline{90-2t}|}+1)$ to $273{\cdot}12a_{\overline{90-2t}|}$. F_t is therefore a discontinuous function and is not strictly suitable for approximate integration, but we may take it as being

$$273{\cdot}12(a_{\overline{90-2t}|}+\tfrac{1}{2}).$$

If we use formula '39*a*', we may take n as being $(102-65)\div 7=5$ say.

t	5	15	25	30	35
$\log (a^{.025}_{\overline{90-2t}|}+\frac{1}{2})$	1·54347	1·49705	1·40829	1·33102	1·20653
$\log v^t_{.04}$	$\bar{1}$·91483	$\bar{1}$·74450	$\bar{1}$·57417	$\bar{1}$·48900	$\bar{1}$·40383
$\log l_{65+t}$	5·46776	5·11956	4·30481	3·54095	2·28711
$\log (l_{65}-l_{65+t})$	4·80872	5·35466	5·52867	5·54962	5·55362
$\log \mu_{65+t}$	$\bar{2}$·69653	$\bar{1}$·07799	$\bar{1}$·44489	$\bar{1}$·64368	$\bar{1}$·87754
log coefft	0·20952	0·34242	0·20952	$\bar{1}$·74819	0·20952
$-2 \log l_{65}$	$\overline{12}$·89228	$\overline{12}$·89228	$\overline{12}$·89228	$\overline{12}$·89228	$\overline{12}$·89228
Total	$\bar{1}$·53311	0·02846	$\bar{1}$·36263	$\bar{2}$·19474	$\bar{3}$·43043
Antilog	·3413	1·0677	0·2305	·0157	·0027

$$\therefore \text{the price} = 2\times 273{\cdot}12(\cdot 3413+1{\cdot}0677+\cdot 2305+\cdot 0157+\cdot 0027)\times 5$$
$$=546{\cdot}24\times 1{\cdot}6579\times 5=\pounds 4528.$$

Rough check

$$\mathring{e}_{\overline{65:65}}=\mathring{e}_{65}+\mathring{e}_{65}-\mathring{e}_{65:65}.$$

Since $a_{65:65}\fallingdotseq a_{73}$ at 2 per cent, this is approximately equal to

$$2\mathring{e}_{65}-\mathring{e}_{73}=16{\cdot}85.$$

Therefore the answer may be expected to be in the neighbourhood of

$$v^{17}_{.04}\times 273a^{\cdot 025}_{\overline{56}|}=\cdot 51337\times 273\times 29{\cdot}9 \text{ approx.}$$
$$=\pounds 4200 \text{ approx.}$$

Example 26·3

A speculator buys two fully-paid whole-life non-profit policies for a total of £13,855. The first policy was effected 27 years ago for a sum assured of £10,000 on a life then aged 50; the second policy was effected 30 years ago for a sum assured of £10,000 on a life then aged 30. Assuming the ultimate mortality of the Hypothetical Select Mortality Table, calculate the probability that the speculation will yield not less than 10 per cent per annum compound interest.

Solution

Denote the lives by (x) and (y). Then the probability is

$$\int_0^n (1 - {}_rp_y)\,{}_tp_x\,\mu_{x+t}\,dt,$$

where $x+n$ is the latest age at which (x) can die and t and r are connected by the relationship

$$13{,}855 = 10{,}000(v^t + v^r) \quad 10 \text{ per cent.}$$

To find n, assume that (y) dies at once, i.e. put $r=0$, when

$$3{,}855 = 10{,}000v^t$$

and $$n = t = \frac{\log \cdot 3855}{\log v} = \frac{-\log \cdot 3855}{\log(1+i)} = \frac{\cdot 4140}{\cdot 0414} = 10.$$

Let us put $x=77$ and $y=60$. Then the required probability

$$= \int_0^{10} (1 - {}_rp_{60})\,{}_tp_{77}\,\mu_{77+t}\,dt$$

$$= \tfrac{5}{3}[u_0 + 4u_5 + u_{10}] \quad \text{by Simpson.}$$

When $t=0$, $r=10$ and when $t=5$, $r=2\cdot 8$. $u_{10}=0$. Therefore the required probability

$$= \frac{5}{3}\left\{\left(1 - \frac{l_{70}}{l_{60}}\right)\mu_{77} + 4\left(1 - \frac{l_{62\cdot 8}}{l_{60}}\right)\frac{l_{82}}{l_{77}}\mu_{82}\right\}$$

$$= \tfrac{5}{3}(\cdot 2737 \times \cdot 09271 + 4 \times \cdot 0589 \times \cdot 5612 \times \cdot 14157)$$

$$= \tfrac{5}{3}(\cdot 0254 + \cdot 0187) = \cdot 0735.$$

As a check, let us put $x=60$ and $y=77$. Then the probability is

$$\tfrac{5}{3}(\cdot 7670 \times \cdot 01868 + 4 \times \cdot 2539 \times \cdot 8855 \times \cdot 03088)$$

$$= \tfrac{5}{3}(\cdot 0143 + \cdot 0278) = \cdot 0702.$$

Hence the probability that the speculation will yield not less than 10 per cent is $\cdot 07$ approximately.

CHAPTER 27

EVALUATION OF FUNCTIONS OF TWO OR MORE LIVES—TRANSFORMATIONS AND OTHER METHODS

27·1. Introduction

In the preceding chapter we have seen how joint-life annuities and contingent assurances can be evaluated, either with the aid of commutation functions or by means of approximate integration. It is now proposed to consider

(1) how joint-life annuities on lives of different ages may be evaluated by reference to tables of single-life annuity values or to tables of values of joint-life annuities on lives of equal age,

(2) the evaluation of contingent assurances by methods which depend upon a law of mortality and by approximate methods other than approximate integration,

(3) the application of Lidstone's approximation to the calculation of joint-life endowment assurance premiums and policy values.

27·2. Joint-life annuities when Gompertz's law holds

When a mortality table follows Gompertz's law,

$$\mu_x = Bc^x$$

for all values of x.

$$\begin{aligned}\therefore \quad {}_np_{xyz\ldots} &= \exp\left\{-\int_0^n (\mu_{x+t}+\mu_{y+t}+\mu_{z+t}+\ldots)\,dt\right\}\\ &= \exp\left\{-\int_0^n Bc^t(c^x+c^y+c^z+\ldots)\,dt\right\}\\ &= \exp\left\{-\int_0^n \mu_{w+t}\,dt\right\},\end{aligned}$$

where
$$\left.\begin{aligned} c^x+c^y+c^z+\ldots &= c^w \\ \text{or}\qquad \mu_x+\mu_y+\mu_z+\ldots &= \mu_w.\end{aligned}\right\} \tag{27·1}$$

It follows that, if w is defined by formula (27·1),

$${}_np_{xyz\ldots} = {}_np_w, \tag{27·2}$$

and $$a_{xyz\ldots} = \sum_{t=1}^{\infty} v^t {}_tp_{xyz\ldots} = \sum_{t=1}^{\infty} v^t {}_tp_w$$

or $$a_{xyz\ldots} = a_w. \qquad (27\cdot3)$$

Hence, if it is desired to obtain $a_{xyz\ldots}$ under a table which follows Gompertz's law, all that is necessary is to find the age w such that

$$\mu_x + \mu_y + \mu_z + \ldots = \mu_w,$$

and then to obtain a_w from a table of single-life annuity values.

27·3. Uniform seniority under Gompertz's law

If only two lives are involved and the table follows Gompertz's law

$$a_{xy} = a_w,$$

where $$c^x + c^y = c^w$$

or $$\mu_x + \mu_y = \mu_w.$$

It should be noted that the substitution of a single life (w) for the joint status xy is made possible by the fact that

$$c^{x+t} + c^{y+t} = c^{w+t}$$

for all values of t. Let us adopt the convention that $x > y$. Then it follows that if we wish to replace the joint-life annuity $a_{x+t:y+t}$ by the single-life annuity a_{w+t}, the amount, $w-x$, by which the age of the substituted life ($w+t$) exceeds the age of the older life ($x+t$) is independent of t and depends only on the difference $x-y$. This is known as the principle of uniform seniority. An expression for $w-x$ in terms of $x-y$ may be obtained as follows:

$$c^w = c^x + c^y.$$

$$\therefore \quad c^{w-x} = 1 + c^{-(x-y)}.$$

$$\therefore \quad w - x = \frac{\log(1 + c^{-(x-y)})}{\log c}. \qquad (27\cdot4)$$

The advantage of the principle of uniform seniority is that a table can be constructed which, if entered with the value of $x-y$, gives the value of $w-x$ immediately, so saving the labour of calculating w by means of the values of μ. As an illustration, the first few lines of the uniform seniority table of the American 1937

Standard Annuity Table are given below. The full table appears in *T.A.S.A.* **39** [1938], 16. By entering a table of this kind with the difference between the ages of (x) and (y) we find the number of years to be added to (x) in order to obtain (w). If three lives are involved, the table must be entered twice, and so on.

TABLE 27·1

1937 Standard Annuity Table. Age of single life corresponding to two joint lives	
Difference in ages (years)	Addition to older age (years)
0	9·122
1	8·632
2	8·160
3	7·707
...	...

In the case of annuitants, it is usual to construct separate mortality tables for males and females. Even if the male and female tables both follow Gompertz's law, an annuity on the joint lives of a male and a female cannot be replaced by a single-life annuity unless the value of c is the same in the two tables, or (to put it another way) unless the mortality assumed for one sex is equivalent to that assumed for the other with a fixed addition to, or deduction from, the age. In the American 1937 Standard Annuity Table the mortality of females is assumed to be equivalent to that of males with a deduction of five years from the age.

When a mortality table is being constructed, it is a great advantage to be able to assume Gompertz's law, since no joint-life functions need be tabulated. Unfortunately, Gompertz's law does not usually give a sufficiently accurate representation of actual mortality except over short ranges of ages. However, it is sometimes used for annuitants' tables, in which young ages are relatively unimportant and in which the advantage of being able to dispense with tables of joint-life annuity values is so great that it may offset the inaccuracy involved in the assumption of Gompertz's law.

27·4. Joint-life annuities when Makeham's law holds

When a mortality table follows Makeham's law,

$$\mu_x = A + Bc^x$$

for all values of x.

$$\therefore \quad {}_np_{xyz\ldots(m)} = \exp\left\{-\int_0^n (\mu_{x+t}+\mu_{y+t}+\mu_{z+t}+\ldots \text{ to } m \text{ terms})\,dt\right\}$$

$$= \exp\left\{-\int_0^n mA + Bc^t(c^x+c^y+c^z+\ldots \text{ to } m \text{ terms})\,dt\right\}$$

$$= \exp\left\{-\int_0^n m\mu_{w+t}\,dt\right\},$$

where $$c^x+c^y+c^z+\ldots \text{ to } m \text{ terms} = mc^w. \qquad (27{\cdot}5)$$

Multiplying both sides of formula (27·5) by B and adding mA to each side, we have

$$(A+Bc^x)+(A+Bc^y)+(A+Bc^z)+\ldots \text{ to } m \text{ terms} = m(A+Bc^w)$$

or $$\mu_x+\mu_y+\mu_z+\ldots \text{ to } m \text{ terms} = m\mu_w. \qquad (27{\cdot}6)$$

It follows that, if w is defined by formula (27·6),

$${}_np_{xyz\ldots(m)} = {}_np_{www\ldots(m)}, \qquad (27{\cdot}7)$$

and $$a_{xyz\ldots(m)} = \sum_{t=1}^{\infty} v^t\,{}_tp_{xyz\ldots(m)} = \sum_{t=1}^{\infty} v^t\,{}_tp_{www\ldots(m)}$$

or $$a_{xyz\ldots(m)} = a_{www\ldots(m)}. \qquad (27{\cdot}8)$$

Hence, if it is desired to obtain $a_{xyz\ldots(m)}$ under a table which follows Makeham's law, and if values of joint-life annuities on m lives of equal age are tabulated, all that is necessary is to find the age w such that

$$\mu_x+\mu_y+\mu_z+\ldots \text{ to } m \text{ terms} = m\mu_w,$$

and then to obtain $a_{www\ldots(m)}$ from the table of joint-life annuity values.

If this method is to apply to a select table it is necessary for $\mu_{[x]+t}$ to be of the form $A_t+B_tc^{x+t}$ during the select period. The effect of this on the form of $l_{[x]+t}$ was investigated in section 13·5 of Volume I.

It should be noted that Makeham's law does not enable us to find an age w such that $a_{\overline{xy}}=a_{\overline{ww}}$. If $a_{\overline{xy}}$ is required, it must be obtained by the relationship

$$\begin{aligned} a_{\overline{xy}} &= a_x + a_y - a_{xy} \\ &= a_x + a_y - a_{ww}. \end{aligned}$$

Example 27·1

Calculate the value of $a_{45:35:30}$ on the basis of the H^{M} Table (Makeham Graduation) at 3 per cent interest.

Solution

$$\begin{aligned} \mu_{45} &= \cdot 01204 \\ \mu_{35} &= \cdot 00854 \\ \mu_{30} &= \underline{\cdot 00768} \\ &\ \underline{\cdot 02826} \div 3 = \cdot 00942 = \mu_{38\cdot 48}. \end{aligned}$$

$$\therefore \quad a_{45:35:30} = a_{38\cdot 48:38\cdot 48:38\cdot 48},$$

whence, by first difference interpolation between the values of $a_{38:38:38}$ and $a_{39:39:39}$,

$$a_{45:35:30} = 12\cdot 035.$$

27·5. Uniform seniority under Makeham's law

Since the substitution of the status ww for the status xy, when Makeham's law holds, depends on the fact that

$$c^{x+t} + c^{y+t} = 2c^{w+t}$$

for all values of t, it follows that the principle of uniform seniority applies. This principle may be enunciated as follows:

If the joint-life annuity $a_{x+t:y+t}$ is replaced by $a_{w+t:w+t}$, and if $x>y$, the amount, $x-w$, by which the age of the older life $(x+t)$ exceeds the age of each substituted life $(w+t)$ is independent of t and depends only on the difference $x-y$.

To obtain an expression for $x-w$ in terms of $x-y$, we have

$$2c^w = c^x + c^y.$$

$$\therefore \quad c^w = \tfrac{1}{2}c^x\{1 + c^{-(x-y)}\}.$$

$$\therefore \quad c^{x-w} = \frac{2}{1 + c^{-(x-y)}}.$$

$$\therefore \quad x - w = \frac{\log 2 - \log\{1 + c^{-(x-y)}\}}{\log c}. \qquad (27\cdot 9)$$

By means of this formula we can construct a table which, if entered with the value of $x-y$, gives the value of $x-w$. The following is a typical example of the first few lines of such a table. The figures are based on a table in *T.S.A.* **3** [1951], 217.

TABLE 27·2

Deduction from older age to obtain two equivalent equal ages	
Difference in ages (years)	Deduction from older age (years)
1	·489
2	·954
3	1·398
4	1·819
...	...

27·6. Another method of evaluating joint-life annuities when Makeham's law holds

An alternative approach to the evaluation of joint-life annuities under Makeham's law is as follows:

$$ {}_nE_{xyz\ldots(m)} = \exp\left\{-\int_0^n (\mu_{x+t}+\mu_{y+t}+\mu_{z+t}+\ldots \text{ to } m \text{ terms}+\delta)\,dt\right\}$$

$$= \exp\left\{-\int_0^n [mA+Bc^t(c^x+c^y+c^z+\ldots \text{ to } m \text{ terms})+\delta]\,dt\right\}.$$

If we make the substitution

$$c^x+c^y+c^z+\ldots \text{ to } m \text{ terms} = c^w, \tag{27·10}$$

we have

$$ {}_nE_{xyz\ldots(m)} = \exp\left\{-\int_0^n [A+Bc^{w+t}+\delta+(m-1)A]\,dt\right\}$$

or

$$ {}_nE_{xyz\ldots(m)} = {}_nE'_w, \tag{27·11}$$

where E' is calculated at a rate of interest i' such that

$$\delta' = \delta+(m-1)A$$

or

$$1+i' = \exp\{\delta+(m-1)A\} = \frac{1+i}{s^{m-1}}. \tag{27·12}$$

Multiplying both sides of formula (27·10) by B and adding mA to each side, we have

$$(A+Bc^x)+(A+Bc^y)+(A+Bc^z)+\dots \text{ to } m \text{ terms}$$
$$=(m-1)A+(A+Bc^w)$$

or $$\mu_x+\mu_y+\mu_z+\dots \text{ to } m \text{ terms}-(m-1)A=\mu_w. \qquad (27\cdot13)$$

It follows that $$a_{xyz\dots(m)}=a'_w, \qquad (27\cdot14)$$

where a' is calculated at rate i', as defined by formula (27·12), and w is defined by formula (27·13).

This method is rarely used because, generally speaking, the method of section 27·4 is more convenient.

*27·7. The general principle of substitution of lives

It is interesting to investigate the general principle underlying the substitution of one life, or two or more lives of equal age, for two or more lives of different ages, and to consider the application of this principle to various laws of mortality.

Since

$${}_nE_{xyz\dots(m)}=\exp\left\{-\int_0^n(\mu_{x+t}+\mu_{y+t}+\mu_{z+t}+\dots \text{ to } m \text{ terms}+\delta)\,dt\right\}$$

and

$${}_nE'_{www\dots(k)}=\exp\left\{-\int_0^n(k\mu_{w+t}+\delta')\,dt\right\},$$

where E' is calculated at a rate of interest i', corresponding to the force of interest δ', it follows that if we can find an age w such that

$$\mu_{x+t}+\mu_{y+t}+\mu_{z+t}+\dots \text{ to } m \text{ terms}+\delta=k\mu_{w+t}+\delta' \qquad (27\cdot15)$$

for all values of t, then we can replace

$$a_{xyz\dots(m)} \quad \text{by} \quad a'_{www\dots(k)},$$

where a' is calculated at rate i'. If $\delta=\delta'$, the rate of interest will be unaltered.

In the case of certain laws of mortality which have been suggested from time to time, it is necessary to make δ' a function of t, say δ_t, but it is not proposed to go into this complication here. We will merely consider a few well-known laws of mortality which permit of the substitution of k lives of equal ages for m lives of different ages without the use of a varying force of interest.

Gompertz's law. $\mu_x = Bc^x$

Equation (27·15) becomes

$$B(c^{x+t}+c^{y+t}+c^{z+t}+\dots \text{ to } m \text{ terms})+\delta = kBc^{w+t}+\delta'.$$

By separating the terms which are independent of t from those which are functions of t we have

$$\delta = \delta' \qquad (27·16)$$

and $$B(c^{x+t}+c^{y+t}+c^{z+t}+\dots \text{ to } m \text{ terms}) = kBc^{w+t}. \qquad (27·17)$$

From formula (27·16) we see that the rate of interest is unaltered. Equation (27·17) has two unknown quantities, k and w, and therefore we can give k any value that we like to choose. It follows that we can replace an annuity on m lives of different ages by an annuity on any number of lives of equal age. It is natural to put $k=1$ so that the substituted annuity is a single-life annuity.

Makeham's law. $\mu_x = A + Bc^x$

Equation (27·15) becomes

$$mA + B(c^{x+t}+c^{y+t}+c^{z+t}+\dots \text{ to } m \text{ terms})+\delta = kA + kBc^{w+t}+\delta'.$$

$$\therefore \quad \delta + (m-k)A = \delta' \qquad (27·18)$$

and $$B(c^{x+t}+c^{y+t}+c^{z+t}+\dots \text{ to } m \text{ terms}) = kBc^{w+t}. \qquad (27·19)$$

We now have two equations, (27·18) and (27·19), with three unknown quantities, k, δ' and w. If we put $k=m$, i.e. if we do not alter the number of lives, it follows from equation (27·18) that $\delta=\delta'$, i.e. the rate of interest is unaltered. We can, however, give k any value that we like to choose, provided that we alter the force of interest from δ to $\delta+(m-k)A$.

Makeham's law applied to select tables. $\mu_{[x]+t} = A_t + B_t c^{x+t}$

Equation (27·15) becomes

$$mA_t + B_t(c^{x+t}+c^{y+t}+c^{z+t}+\dots \text{ to } m \text{ terms})+\delta$$
$$= kA_t + kB_t c^{w+t}+\delta'.$$

By equating terms which are independent of t, terms which are functions of A_t and terms which are functions of $B_t c^t$, we have

$$\delta = \delta', \tag{27·20}$$

$$mA_t = kA_t \tag{27·21}$$

and $$B_t(c^{x+t} + c^{y+t} + c^{z+t} + \ldots \text{ to } m \text{ terms}) = kB_t c^{w+t}, \tag{27·22}$$

i.e. three equations with three unknown quantities.

It follows from equations (27·20) and (27·21) that the rate of interest and the number of lives are unaltered; w is obtained from equation (27·22).

Makeham's second modification of Gompertz's law. $\mu_x = A + Hx + Bc^x$

Equation (27·15) becomes

$$\begin{aligned} &mA + H(x+y+z+\ldots \text{ to } m \text{ terms}) \\ &\qquad + mHt + B(c^{x+t} + c^{y+t} + c^{z+t} + \ldots \text{ to } m \text{ terms}) + \delta \\ &\qquad\qquad = kA + kHw + kHt + kBc^{w+t} + \delta'. \end{aligned}$$

By equating like terms, we have

$$\delta + (m-k)A + H(x+y+z+\ldots \text{ to } m \text{ terms}) - kHw = \delta', \tag{27·23}$$

$$mHt = kHt \tag{27·24}$$

and $$B(c^{x+t} + c^{y+t} + c^{z+t} + \ldots \text{ to } m \text{ terms}) = kBc^{w+t}. \tag{27·25}$$

From equation (27·24), $k = m$. Hence w may be obtained from equation (27·25), and finally from equation (27·23) we may obtain δ', which is equal to $\delta + H(x+y+z+\ldots \text{ to } m \text{ terms} - mw)$. Thus we cannot alter the number of lives and we must alter the rate of interest.

The double geometric law

This is usually written in the form

$$\mu_x = A + Bc^x + Mn^x,$$

but for the present purpose it is convenient to write

$$\mu_x = A + B_1 c_1^x + B_2 c_2^x.$$

Equation (27·15) becomes

$$mA+B_1(c_1^{x+t}+c_1^{y+t}+c_1^{z+t}+\dots \text{ to } m \text{ terms})$$
$$+B_2(c_2^{x+t}+c_2^{y+t}+c_2^{z+t}+\dots \text{ to } m \text{ terms})+\delta$$
$$=kA+kB_1c_1^{w+t}+kB_2c_2^{w+t}+\delta'.$$

By equating like terms, we have

$$\delta+(m-k)A=\delta', \qquad (27{\cdot}26)$$

$$B_1(c_1^{x+t}+c_1^{y+t}+c_1^{z+t}+\dots \text{ to } m \text{ terms})=kB_1c_1^{w+t} \qquad (27{\cdot}27)$$

and $$B_2(c_2^{x+t}+c_2^{y+t}+c_2^{z+t}+\dots \text{ to } m \text{ terms})=kB_2c_2^{w+t}. \qquad (27{\cdot}28)$$

From equations (27·27) and (27·28)

$$\frac{c_1^w}{c_2^w}=\frac{c_1^x+c_1^y+c_1^z+\dots \text{ to } m \text{ terms}}{c_2^x+c_2^y+c_2^z+\dots \text{ to } m \text{ terms}},$$

whence w can be obtained. We can then obtain k from either formula (27·27) or (27·28) and finally δ' from formula (27·26).

It will be noted that it is necessary to alter the number of lives as well as the rate of interest. The number of substituted lives will not, generally speaking, be an integer.

The double geometric law. An alternative procedure

We may regard the force of mortality as being composed of two partial forces, ${}^1\mu_x=A+B_1c_1^x$ and ${}^2\mu_x=B_2c_2^x$, the ages of the substituted lives being w_1 for the force ${}^1\mu_x$ and w_2 for the force ${}^2\mu_x$. We can then put $k=m$ (whence $\delta=\delta'$) and the expressions on the right-hand side of equations (27·27) and (27·28) become $mB_1c_1^{w_1+t}$ and $mB_2c_2^{w_2+t}$, respectively. Therefore

$$mc_1^{w_1}=c_1^x+c_1^y+c_1^z+\dots \text{ to } m \text{ terms}$$

and $$mc_2^{w_2}=c_2^x+c_2^y+c_2^z+\dots \text{ to } m \text{ terms}.$$

Hence w_1 and w_2 can be calculated.

It is found that for the values of c_1 and c_2 which occur in practice w_2-w_1 has a very limited range. If a table of annuity values is prepared corresponding to integral values of w_1 and w_2-w_1, any required annuity value can be obtained by interpolation.

This method has the advantage that neither the rate of interest nor the number of lives is altered.

The triple geometric law. $\mu_x = A + B_1 c_1^x + B_2 c_2^x + B_3 c_3^x$

The alternative procedure followed above for the double geometric law can be extended to the triple geometric law. We shall have

$$mc_1^{w_1} = c_1^x + c_1^y + c_1^z + \dots \text{ to } m \text{ terms},$$

$$mc_2^{w_2} = c_2^x + c_2^y + c_2^z + \dots \text{ to } m \text{ terms}$$

and

$$mc_3^{w_3} = c_3^x + c_3^y + c_3^z + \dots \text{ to } m \text{ terms.}$$

When only two lives are involved there is only one possible value of $w_3 - w_2$ for any given value of $w_2 - w_1$. Accordingly it will not be necessary to calculate the value of w_3 when any individual value of a_{xy} is required. A prepared table of annuity values similar to that which is used for the double geometric law can be constructed; it will not be necessary to allow for variations in $w_3 - w_2$ when preparing the table but of course the appropriate value of w_3 must be taken into account.

When more than two lives are involved, the variations in $w_3 - w_2$ for any given value of $w_2 - w_1$ are, in practice, very small and a method of interpolation can still be devised. (See H. O'Brien and J. A. Cameron, *J.I.A.* **68** [1937], 541.)

27·8. The substitution of a single life when Gompertz's law does not hold

We have seen that if a mortality table follows Gompertz's law, we can write

$$a_{xy} = a_w$$

and we can obtain w from the relationship

$$\mu_x + \mu_y = \mu_w.$$

We could equally well obtain w by writing

$$\mu_{x+t} + \mu_{y+t} = \mu_{w+t},$$

where t could have any value.

If the table does not follow Gompertz's law, there must still be an age w such that

$$a_{xy} = a_w,$$

but there is no ready method by which w can be found accurately. By writing

$$\mu_{x+t} + \mu_{y+t} = \mu_{w+t}$$

and giving t various values, we may obtain different values of w but we have no means of knowing which of these is the correct one.

An empirical method, suggested by Elderton and A. E. King in *J.I.A.* **44** [1910], 293, is to write

$$\mu_{x+n} + \mu_{y+n} = \mu_{w+n},$$

where n is approximately equal to the value of the annuity.

This method is not very accurate but it can be applied quite rapidly and it is useful when a close approximation is not required.

Example 27·2

By substituting a single life, calculate the value of $a_{42:36}$ approximately on the basis of the A 1924–29 Table (ultimate mortality) with interest at $2\frac{1}{2}$ per cent per annum.

Solution

To obtain a rough idea of the value of the annuity (and hence of n) we may try putting $t=15$ and $t=20$.

$$\begin{aligned} \mu_{42+15} &= \cdot 01387 \qquad & \mu_{42+20} &= \cdot 02308 \\ \mu_{36+15} &= \cdot 00800 & \mu_{36+20} &= \cdot 01256 \\ & \quad \cdot 02187 & & \quad \cdot 03564 \\ &= \mu_{46+15} \text{ approx.} & &= \mu_{46+20} \text{ approx.} \end{aligned}$$

$\therefore \; a_{42:36} \doteqdot a_{46} = 18$ to the nearest integer.

If we now put $n=18$, we have

$$\begin{aligned} \mu_{42+18} &= \cdot 01894 \\ \mu_{36+18} &= \cdot 01040 \\ & \quad \cdot 02934 = \mu_{46\cdot 5+18} \text{ approx.} \end{aligned}$$

$$\therefore \quad a_{42:36} \doteqdot a_{46\cdot 5} = 18\cdot 052.$$

The true value, as shown in the A 1924–29 Monetary Tables (Joint Lives) is 17·824.

A similar method may be used when a joint-life temporary annuity is required. For this purpose n may be taken as being approximately one-third of the term of the annuity. If the term is short, a fairly close approximation will usually be obtained.

Example 27·3

By substituting a single life, calculate the value of $\ddot{a}_{51:48:\overline{10}|}$ approximately on the basis of the A 1924–29 Table (ultimate mortality) with interest at $2\frac{1}{2}$ per cent per annum.

Solution

$$\mu_{51+3} = \cdot 01040$$
$$\mu_{48+3} = \cdot 00800$$
$$\cdot 01840 = \mu_{56\cdot 71+3}.$$

Therefore $$\ddot{a}_{51:48:\overline{10}|} \doteq \ddot{a}_{56\cdot 71:\overline{10}|} = 8\cdot 282.$$

The true value is

$$\ddot{a}_{51:48} - \frac{D_{60}}{D_{51}} \frac{l_{57}}{l_{48}} a_{60:57} = 8\cdot 280.$$

Thomas Simpson in 1752 suggested a method of evaluating a_{xyz} when only one-life and two-life annuities are available. Let (x) be the youngest life. Then the method consists in finding w such that $a_{yz} = a_w$ and writing

$$a_{xyz} = a_{xw}.$$

It has been found that this method usually gives too large a value.

27·9. The substitution of lives of equal age when Makeham's law does not hold

When a table does not follow Gompertz's or Makeham's law, there must still be an age w such that

$$a_{xy} = a_{ww},$$

but there is no method by which w can be found accurately.

Various approximate methods of obtaining w are available. For example:

(1) A table of uniform seniority which has been constructed in connexion with a mortality table which follows Makeham's law may be used.

(2) If it is expected that a number of annuity values will be required from time to time and it is not desired to construct a table, on the lines suggested in Chapter 26, giving values of a_{xy} for every

second, third or fifth age, a few specimen values of a_{xy} for various values of x and y may be calculated and a table of uniform seniority, designed to reproduce these values approximately, may be constructed.

(3) A method similar to that of section 27·8 may be used, w being obtained from the relationship

$$\mu_{x+n}+\mu_{y+n}=2\mu_{w+n},$$

where n is approximately equal to the value of the annuity.

It is an obvious convenience for a life office which transacts joint-life assurance business to be able to include in its prospectus a concise table by means of which members of the public can easily calculate the rate of premium for any combination of ages x and y. Some offices solve this problem by publishing rates for lives of equal age and providing a table of uniform seniority, in a simplified form, which gives the ages of the substituted lives in integers only. The following is a typical table.

TABLE 27·3

Difference in ages next birthday of the two lives	Deduction from the age next birthday of the older life to obtain the age next birthday of the equivalent lives
1	0
2– 3	1
4– 5	2
6– 8	3
9–12	4
13–17	5
18–24	6
25 and over	7

The use of a table of this kind involves a slight theoretical inaccuracy, especially if the table does not follow Makeham's law. The method is also open to the practical objection that an alteration in the age next birthday of the younger life often has no effect on the premium. However, it may be considered that these disadvantages are more than offset by the usefulness of the table.

*27·10. Contingent assurances when Gompertz's law holds

When a mortality table follows Gompertz's law,

$$\bar{A}^{1}_{xy} = \int_0^\infty v^t {}_tp_{xy} Bc^{x+t}\,dt$$

$$= \frac{c^x}{c^x + c^y} \int_0^\infty v^t {}_tp_{xy} Bc^t(c^x + c^y)\,dt$$

$$= \frac{\mu_x}{\mu_x + \mu_y} \int_0^\infty v^t {}_tp_{xy}(\mu_{x+t} + \mu_{y+t})\,dt.$$

$$\left.\begin{aligned} \therefore \quad \bar{A}^{1}_{xy} &= \frac{\mu_x}{\mu_x + \mu_y} \bar{A}_{xy} \\ &= \frac{\mu_x}{\mu_w} \bar{A}_w, \end{aligned}\right\} \qquad (27\cdot29)$$

where $\mu_x + \mu_y = \mu_w$.

It follows that

$$\left.\begin{aligned} P(\bar{A}^{1}_{xy}) &= \frac{\mu_x}{\mu_x + \mu_y} P(\bar{A}_{xy}) \\ &= \frac{\mu_x}{\mu_w} P(\bar{A}_w). \end{aligned}\right\} \qquad (27\cdot30)$$

*27·11. Contingent assurances when Makeham's law holds

Hume and Stott in *J.I.A.* **39** [1905], 370, suggested the following method of evaluating contingent assurances under a table which follows Makeham's law.

$$\bar{A}^{1}_{xy} = \int_0^\infty v^t {}_tp_{xy}(A + Bc^{x+t})\,dt$$

$$= A\bar{a}_{xy} + \frac{c^x}{c^x + c^y} \int_0^\infty v^t {}_tp_{xy} Bc^t(c^x + c^y)\,dt$$

$$= A\bar{a}_{ww} + \frac{c^x}{2c^w} \int_0^\infty v^t {}_tp_{ww} 2Bc^{w+t}\,dt,$$

where $c^x + c^y = 2c^w$, or $\mu_x + \mu_y = 2\mu_w$.

$$\therefore \quad \bar{A}^{1}_{xy} = A\bar{a}_{ww} + \frac{c^x}{2c^w} \int_0^\infty v^t {}_tp_{ww}(2\mu_{w+t} - 2A)\,dt$$

$$= A\bar{a}_{ww} + \frac{c^x}{2c^w}(\bar{A}_{ww} - 2A\bar{a}_{ww}).$$

This may be written

$$\bar{A}^1_{xy} = c^x \frac{\frac{1}{2}\bar{A}_{ww} - A\bar{a}_{ww}}{c^w} + A\bar{a}_{ww} \qquad (27\cdot31)$$

or

$$\bar{A}^1_{xy} = c^x f_1(ww) + f_2(ww), \qquad (27\cdot32)$$

where

$$f_1(ww) = \frac{\frac{1}{2}\bar{A}_{ww} - A\bar{a}_{ww}}{c^w}$$

and

$$f_2(ww) = A\bar{a}_{ww}.$$

By means of formula (27·32), $\bar{A}^1_{xy}$ can be evaluated rapidly with the aid of prepared tables of c^x, $f_1(ww)$ and $f_2(ww)$.

The method can be readily extended to give a contingent assurance payable on the first death of any number of lives. For example,

$$\bar{A}^1_{xyz} = c^x \frac{\frac{1}{3}\bar{A}_{www} - A\bar{a}_{www}}{c^w} + A\bar{a}_{www}. \qquad (27\cdot33)$$

An alternative method, due to A. W. Evans, *J.I.A.* **56** [1925], 220, is as follows:

By the theorem of mean value

$$\int_0^\infty v^t {}_tp_{xy}\, \mu_{x+t}\, dt = \mu_{x+n} \int_0^\infty v^t {}_tp_{xy}\, dt, \qquad (27\cdot34)$$

where μ_{x+n} is a mean value of μ_{x+t}. Similarly,

$$\int_0^\infty v^t {}_tp_{xy} (\mu_{x+t} + \mu_{y+t})\, dt = (\mu_{x+n} + \mu_{y+n}) \int_0^\infty v^t {}_tp_{xy}\, dt. \qquad (27\cdot35)$$

If the mortality table does not follow Makeham's law, the value of n which satisfies equation (27·34) will not necessarily satisfy equation (27·35). However, if the table follows Makeham's law, both equations reduce to

$$\int_0^\infty v^t {}_tp_{xy}\, c^t\, dt = c^n \int_0^\infty v^t {}_tp_{xy}\, dt,$$

and it follows that the same value of n satisfies both equations.

But formula (27·34) can be written

$$\bar{A}^1_{xy} = \mu_{x+n}\, \bar{a}_{xy} = \mu_{x+n}\, \bar{a}_{ww} \qquad (27\cdot36)$$

(where $\mu_x + \mu_y = 2\mu_w$) and formula (27·35) can be written

$$\mu_{x+n} + \mu_{y+n} = \bar{P}(\bar{A}_{xy})$$

or

$$\mu_{w+n} = \tfrac{1}{2}\bar{P}(\bar{A}_{ww}). \qquad (27\cdot37)$$

Therefore, if we calculate the value of n which satisfies equation (27·37) we can obtain $\bar{A}^1_{xy}$ from formula (27·36). To facilitate the calculation it is useful to have a prepared table of n. In the absence of such a table $w+n$ (and hence n) can be obtained by entering a table of μ inversely with the value of $\frac{1}{2}\bar{P}(\bar{A}_{ww})$.

Evans's method can be readily extended to three or more lives. Thus

$$\bar{A}^1_{xyz}=\mu_{x+n}\bar{a}_{www},$$

where n is such that

$$\mu_{w+n}=\tfrac{1}{3}\bar{P}(\bar{A}_{www}).$$

It is interesting to note that since the same value of n satisfies the two equations

$$\bar{A}^1_{xy}=\mu_{x+n}\bar{a}_{xy}$$

and

$$\bar{A}_{xy}=(\mu_{x+n}+\mu_{y+n})\bar{a}_{xy},$$

it follows that

$$\left.\begin{aligned}\bar{A}^1_{xy}&=\frac{\mu_{x+n}}{\mu_{x+n}+\mu_{y+n}}\bar{A}_{xy}\\&=\frac{\mu_{x+n}}{2\mu_{w+n}}\bar{A}_{ww},\end{aligned}\right\}\qquad(27\cdot38)$$

and that

$$\left.\begin{aligned}P(\bar{A}^1_{xy})&=\frac{\mu_{x+n}}{\mu_{x+n}+\mu_{y+n}}P(\bar{A}_{xy})\\&=\frac{\mu_{x+n}}{2\mu_{w+n}}P(\bar{A}_{ww}).\end{aligned}\right\}\qquad(27\cdot39)$$

The resemblance between these formulae and formulae (27·29) and (27·30) should be noted.

Neither Hume and Stott's method nor Evans's method is applicable to select tables. See, however, the remarks on this subject in section 27·13.

*27·12. Contingent assurances—approximate methods based on Gompertz's and Makeham's laws

When a mortality table does not follow either Gompertz's or Makeham's law, various approximate methods (in addition to approximate integration) are available for evaluating contingent assurances.

Elderton and King, in their paper already referred to in section 27·8, suggested the following empirical method:

First obtain w by means of the relationship

$$\mu_{x+n}+\mu_{y+n}=\mu_{w+n},$$

where n is approximately equal to a_y. Then

$$\bar{A}^{1}_{xy} \doteq \frac{\mu_{x+n}}{\mu_{w+n}} \bar{A}_{w}.$$

It will be seen that this is a modification of formula (27·29), just as the method of section 27·8 is a modification of formula (27·3). It is very convenient for obtaining a rough approximation with a minimum of arithmetical work and it is suitable if joint-life annuities are not tabulated.

Evans's method may be adapted as follows for use with a table which does not follow Makeham's law:

Let us assume that the values of n which satisfy the two equations

$$\bar{A}^{1}_{xy} = \mu_{x+n}\bar{a}_{xy}$$

and

$$\bar{A}_{xy} = (\mu_{x+n} + \mu_{y+n})\bar{a}_{xy}$$

are approximately equal. Then it follows that

$$\bar{A}^{1}_{xy} \doteq \frac{\mu_{x+n}}{\mu_{x+n} + \mu_{y+n}} \bar{A}_{xy},$$

where n is such that $\mu_{x+n} + \mu_{y+n} = \bar{P}(\bar{A}_{xy})$.

This method is not suitable unless joint-life annuities are available.

27·13. Contingent assurances—other approximate methods

An approximation for $\bar{A}^{1}_{xy}$ which has been found useful in practice is based on the formula

$$\mu_{x+t} \doteq \frac{l_{x+t-1} - l_{x+t+1}}{2l_{x+t}}.$$

$$\bar{A}^{1}_{xy} = \int_0^\infty v^t {}_tp_{xy}\,\mu_{x+t}\,dt$$

$$\doteq \int_0^\infty v^t {}_tp_{xy} \frac{l_{x+t-1} - l_{x+t+1}}{2l_{x+t}}\,dt$$

$$= \frac{1}{2}\left\{\frac{l_{x-1}}{l_x}\int_0^\infty v^t \frac{l_{x+t-1}}{l_{x-1}} {}_tp_y\,dt - \frac{l_{x+1}}{l_x}\int_0^\infty v^t \frac{l_{x+t+1}}{l_{x+1}} {}_tp_y\,dt\right\}.$$

$$\therefore \quad \bar{A}^{1}_{xy} \doteq \frac{1}{2}\left\{\frac{\bar{a}_{x-1:y}}{p_{x-1}} - p_x\bar{a}_{x+1:y}\right\} \qquad (27·40)$$

$$\doteq \frac{1}{2}\left\{\frac{a_{x-1:y}}{p_{x-1}} - p_x a_{x+1:y}\right\}. \qquad (27·41)$$

Another approximation is obtained by differentiating a_{xy} with respect to x, keeping y constant.

$$\frac{d}{dx}\bar{a}_{xy}=\frac{d}{dx}\int_0^\infty v^t\,{}_tp_{xy}\,dt$$

$$=\int_0^\infty v^t\,{}_tp_y\frac{d}{dx}\left(\frac{l_{x+t}}{l_x}\right)dt$$

$$=\int_0^\infty v^t\,{}_tp_y\frac{(l_x\mu_x)\,l_{x+t}-(l_{x+t}\mu_{x+t})\,l_x}{l_x^2}\,dt$$

$$=\int_0^\infty v^t\,{}_tp_{xy}(\mu_x-\mu_{x+t})\,dt$$

$$=\mu_x\bar{a}_{xy}-\bar{A}^1_{xy}.$$

$$\therefore\quad \bar{A}^1_{xy}=\mu_x\bar{a}_{xy}-\frac{d}{dx}\bar{a}_{xy}$$

$$\risingdotseq\mu_x\bar{a}_{xy}+\tfrac{1}{2}(\bar{a}_{x-1:y}-\bar{a}_{x+1:y}) \qquad (27\cdot42)$$

$$\risingdotseq\mu_x(a_{xy}+\tfrac{1}{2})+\tfrac{1}{2}(a_{x-1:y}-a_{x+1:y}). \qquad (27\cdot43)$$

The following table shows the value of $\bar{A}^1_{30:45}$ on the basis of A 1924–29 ultimate mortality and $2\frac{1}{2}$ per cent interest according to various formulae.

Formula number	Formula	Value
—	$\bar{M}^1_{xy}/D_{xy}$	·11554
(26·20)	Hardy's '39*a*'	·11669
(26·26)	Beard's (91)	·11622
(27·41)	$\frac{1}{2}\left\{\frac{a_{x-1:y}}{p_{x-1}}-p_x a_{x+1:y}\right\}$	·116
(27·43)	$\mu_x(a_{xy}+\frac{1}{2})+\frac{1}{2}(a_{x-1:y}-a_{x+1:y})$	·116

In evaluating formulae (27·41) and (27·43), $a_{29:45}$ and $a_{31:45}$ were obtained by first-difference interpolation between the tabulated values $a_{27:45}$, $a_{30:45}$ and $a_{33:45}$. In view of the nature of the approximations used in developing the formulae, refinements in the calculation of the annuity values would be out of place.

Formulae (27·40) to (27·43) are not suitable for use with select mortality tables, because they would involve meaningless expressions like $a_{[x]-1:[y]}$. Other approximations can be produced which avoid the suffix $[x]-1$ but these will contain expressions like $a_{[x]+1:[y]}$, which are not normally tabulated. Accordingly, if the value of $\bar{A}^{\,1}_{[x]:[y]}$ is required it will usually be necessary to resort to approximate integration. However, with modern mortality tables it is often found that the error involved in using ultimate mortality for both $[x]$ and $[y]$ is very small. Even if $\bar{A}^{\,1}_{[x]:y}$ or $\bar{A}^{\,1}_{x:[y]}$ is required, a sufficiently accurate result for practical purposes will often be obtained by using ultimate mortality for both lives. This is illustrated by the following table, which is again based on A 1924–29 mortality and $2\frac{1}{2}$ per cent interest. The values have been calculated by means of commutation functions.

Function	Value
$\bar{A}^{\,1}_{[30]:[45]}$	·11478
$\bar{A}^{\,1}_{[30]:45}$	·11437
$\bar{A}^{\,1}_{30:[45]}$	·11595
$\bar{A}^{\,1}_{30:45}$	·11554

*27·14. Contingent survivorship annuities

Various approximate methods for evaluating contingent survivorship annuities without recourse to approximate integration have been developed from time to time. These methods are based on the relationship

$$\bar{a}^{\,1}_{yz|x}=\int_0^\infty v^t\,{}_tp_{xyz}\,\mu_{y+t}\,\bar{a}_{x+t}\,dt$$

$$=\int_0^\infty \frac{\mu_{y+t}}{\mu_{y+t}+\mu_{z+t}}\,v^t\,{}_tp_{xyz}(\mu_{y+t}+\mu_{z+t})\,\bar{a}_{x+t}\,dt$$

$$=\frac{\mu_{y+n}}{\mu_{y+n}+\mu_{z+n}}\,\bar{a}_{yz|x},$$

where $\dfrac{\mu_{y+n}}{\mu_{y+n}+\mu_{z+n}}$ is a mean value of $\dfrac{\mu_{y+t}}{\mu_{y+t}+\mu_{z+t}}$.

The reader who is interested in this subject is referred to the paper by Elderton and King, already mentioned in this chapter, and to papers by A. W. Evans, *J.I.A.* **56** [1925], 220 and V. G. C. Callaway, *J.I.A.* **63** [1932], 67.

On the rare occasions when contingent survivorship annuities are required, they may conveniently be evaluated by approximate integration.

27·15. Lidstone's approximation for joint-life endowment assurance premiums

Lidstone's approximate formula,

$$P_{xy:\overline{n}|} \doteqdot P_{x:\overline{n}|} + P_{y:\overline{n}|} - P_{\overline{n}|},$$

which was demonstrated in section 8 of Chapter 23 and numbered (23·19), is useful for the approximate evaluation of net premiums for joint-life endowment assurances when joint-life functions are not available. Similarly, the approximate formulae given in section 9 of the same chapter may be used for evaluating net premiums for joint-life endowment assurances on three or four lives.

The following table illustrates the degree of accuracy of these approximations:

Function	Formula	Approximate value	True value
$P_{40:40:\overline{20}|}$	(23·19)	·0449	·0449
$P_{40:40:40:\overline{20}|}$	(23·22)	·0482	·0483
$P_{40:40:40:\overline{20}|}$	(23·24)	·0482	·0483
$P_{40:40:40:\overline{20}|}$	(23·25)	·0483	·0483
$P_{40:40:40:\overline{20}|}$	(23·26)	·0483	·0483
$P_{40:40:40:40:\overline{20}|}$	(23·23)	·0515	·0518

All figures are based on A 1924–29 ultimate mortality with $2\frac{1}{2}$ per cent interest. Equal ages have been used because accurate values for three and four lives of different ages are not available.

These formulae can also be used for obtaining approximately the policy value of a joint-life endowment assurance. For example, since

$$P_{xy:\overline{n}|} \doteqdot P_{x:\overline{n}|} + P_{y:\overline{n}|} - P_{\overline{n}|},$$

it follows that

$${}_tV_{xy:\overline{n}|} \doteqdot {}_tV_{x:\overline{n}|} + {}_tV_{y:\overline{n}|} - {}_tV_{\overline{n}|}.$$

EXAMPLES

Examples 27·1, 27·2 and 27·3 will be found in the text of the chapter.

Example 27·4

A mortality table follows Makeham's Law with constants g, s and c', where

$$\log c' = 1{\cdot}25 \log c$$

and g, s and c are Makeham constants of a Standard Table ($\log_e s = -{\cdot}006192$).

Show that the value of $\bar{a}_{xyzuv:\overline{n}|}$ on the special table is equal to the value of a similar five-life function on the Standard Table at a different rate of interest. Find the Standard Table function and the rate of interest corresponding to the special table function at 4 per cent when $x=y=z=40$, $u=v=60$ and $n=20$. You are given that $\delta_{.04}={\cdot}039221$.

Solution

$$\begin{aligned}\bar{a}'_{xyzuv:\overline{n}|} &= \int_0^n v^t\, {}_tp_{xyzuv}\, dt \\ &= \int_0^n v^t s^{5t} g^{(c^{1\cdot25x}+c^{1\cdot25y}+\ldots)\,(c^{1\cdot25t}-1)}\, dt \\ &= {\cdot}8\int_0^{1\cdot25n} v^t s^{5t} g^{(c^{1\cdot25x}+c^{1\cdot25y}+\ldots)(c^{1\cdot25t}-1)}\, d(1{\cdot}25t) \\ &= {\cdot}8\int_0^{1\cdot25n} v^{\cdot8t} s^{4t} g^{(c^{1\cdot25x}+c^{1\cdot25y}+\ldots)(c^t-1)}\, dt \\ &= {\cdot}8\bar{a}'_{1\cdot25x:1\cdot25y:\ldots\overline{1\cdot25n}|},\end{aligned}$$

$\bar{a}'$ being calculated at a rate of interest i' such that

$$v'^t s^{5t} = v^{\cdot8t} s^{4t}$$

or

$$\delta' - \log_e s = {\cdot}8\delta.$$

In the example given in the question, we require

$${\cdot}8\bar{a}'_{50:50:50:75:75:\overline{25}|}$$

at rate i' such that $\delta' = \cdot 8\delta_{\cdot 04} + \log_e s$

$$= \cdot 031377 - \cdot 006192$$

$$= \cdot 025185$$

or $i' = \cdot 0255$ approx.

Example 27·5

Develop a formula for the probability that (x) will die first of three lives (x), (y) and (z) where Makeham's law holds.

Solution

It was shown in section 27·11 that

$$\bar{A}^1_{xyz} = c^x \frac{\frac{1}{3}\bar{A}_{www} - A\bar{a}_{www}}{c^w} + A\bar{a}_{www}.$$

The corresponding formula for ${}_\infty q^1_{xyz}$ is obtained by making the rate of interest zero in the above expression.

When the rate of interest is zero

$$\bar{A}_{www} = 1$$

and $$\bar{a}_{www} = \mathring{e}_{www}.$$

Hence $${}_\infty q^1_{xyz} = c^x \frac{\frac{1}{3} - A\mathring{e}_{www}}{c^w} + A\mathring{e}_{www}.$$

When $x = y = z$, $c^x = c^w$ and this expression reduces to $\frac{1}{3}$ which is otherwise obvious.

CHAPTER 28

PENSION FUNDS—SERVICE TABLES, SALARY SCALES AND VALUATION OF CONTRIBUTIONS

28·1. Introduction

Consideration will be given in this and the next chapter to staff pension funds which provide, in return for contributions depending on salary, pensions which also depend on salary. Membership of such funds is confined to persons who are in the service of a specified employer or group of employers and it ceases when they leave that service.

The main purpose of a pension fund is to provide pensions to the members on their retirement; most funds provide a benefit (usually a cash payment equal to the member's contribution with or without interest) on death before retirement or on withdrawal. Some pension funds provide additional benefits in the form of pensions to widows and orphans, but these benefits usually form the subject of a separate fund set up for the purpose; they will be considered in Chapter 30.

Different pension schemes vary considerably as regards the relationship between contributions and salary, the precise nature of the benefits, the conditions which have to be fulfilled in order that a member may be entitled to benefit, and so on. In recent years there has been a tendency for schemes to become standardized and the formulae which will be developed are suitable for the types of scheme which are most popular in Great Britain at the present time. They should, however, be regarded as illustrative of the principles to be adopted in valuation rather than as necessarily being suitable for use without modification. Where the conditions of a scheme do not conform to those assumed there should be no difficulty in producing appropriate formulae by similar methods.

It is usual for both the employer and the employee to contribute to the fund. The employees' contributions are payable by deduction from their salary, which may be paid either weekly or monthly. For valuation purposes, contributions are assumed to be payable continuously; the same assumption is made in respect of pension payments. The benefit on death or withdrawal is assumed to be payable at the moment when the death or withdrawal occurs.

For calculations in respect of a pension scheme a 'service table' is used. This is a multiple-decrement table which is suited to the circumstances of the particular scheme. The decrements which normally occur in a pension scheme, and which will be considered here, are:

(i) withdrawal from the service, denoted by w,
(ii) death within the service, denoted by d,
(iii) ill-health retirement, denoted by i,
(iv) age retirement, denoted by r.

If the benefit payable on voluntary resignation differs from that payable on dismissal, it may be necessary to treat these two types of withdrawal as two separate decrements. The word 'retirement' is understood to mean 'retirement under conditions which entitle the member to a pension in accordance with the rules of the scheme'. 'Ill-health retirement' means 'retirement as a result of ill-health'; 'age retirement' means 'retirement which takes place, not as a result of ill-health, but by reason of age or length of service'. If the rules of the scheme provide that the amount of the pension depends on whether the retirement is due to 'ill-health' or to 'age', it will be necessary to have two separate factors for valuing the 'ill-health' and 'age' pensions. If the scheme does not differentiate between the two types of retirement, the two factors can be combined in one.

It will be assumed, purely for convenience, that the rules of the scheme permit of age retirements at any time between exact age 60 and exact age 65 and that members who are still in the service when they attain age 65 are required to retire immediately. A service table for use in connexion with such a scheme will contain ill-health retirements up to age 64 inclusive and age retirements from age 60 to age 65 both inclusive. It will also be assumed that all forces of decrement are continuous throughout the range of ages over which they operate, except that the force of age retirement is discontinuous at ages 60 and 65. An example of such a table is given on page 247.

It is often found that the rates of withdrawal are a function of the duration of service as well as the age. Where this feature is marked and entrants are not limited to the younger ages it may be desirable to use select rates of withdrawal. Owing to the additional work involved, the use of select withdrawal rates is normally justified only when the fund is large and punched-card equipment is available for the calculations. The development of formulae on a select basis presents no particular difficulty and will not be considered here.

For calculations which depend upon mortality after retirement it is necessary to have a mortality table (or at least a table of annuity values) applicable to pensioners. Two different tables may be needed—one for ill-health pensioners and one for age pensioners.

A scale of salary increases also will be needed. The nature and purpose of this scale will be described in section 3 of this chapter.

In practical work it is necessary to distinguish between four different categories of employee which may be referred to as male clerical staff, female clerical staff, male workpeople and female workpeople, respectively. A service table and salary scale which are suitable for one of these categories will probably not be satisfactory for any other; accordingly the four categories have to be kept separate.

We shall not be concerned with the problems arising in connexion with the choice of suitable assumptions as to mortality, withdrawal, retirement or increases of salary. It will be assumed that these matters have been settled and attention will be confined to the derivation of formulae for valuing the contributions and benefits on the basis of a given service table, salary scale and pensioners' mortality table.

The contributions to a pension fund are usually a percentage of the salary. This percentage may be the same for all members or it may vary with the age at entry. When a pension fund is formed it is necessary to fix a suitable rate or rates of contribution in order that the fund may be in a position to grant the desired benefits.

The rules of many pension schemes provide for valuations at quinquennial intervals. The financial year of most firms ends on

31 December and this is the date which is usually chosen for the periodical valuations. In this and the succeeding chapter, the problem of valuing the contributions and benefits will be approached on the assumption that a valuation of an established fund is being made, so that the rates of contribution have already been fixed. Except where otherwise stated, it will be assumed that the valuation date is 31 December. The discussion of the determination of the rates of contribution will be deferred until after the formulae for valuing the contributions and benefits have been derived.

The principle adopted will be to write down expressions which are, as nearly as possible, theoretically accurate for the values of the contributions and benefits, and to introduce such approximations as may be necessary in order to reduce the expressions to a form in which they can be most conveniently evaluated. The reason for adopting this approach is that it makes clear the theoretical foundations on which the formulae rest. It is possible to arrive at the same formulae by methods which have a simpler appearance but are less precise.

28·2. Notation

In practical pension fund work it is not usual to employ the notation described in Chapter 19 for a multiple-decrement table. A simpler notation is commonly used. For convenience the two notations are shown side by side here:

TABLE 28·1

Multiple-decrement table notation	Notation used in pension fund work
$(al)_x$	l_x
$(ad)_x^w$	w_x
$(ad)_x^d$	d_x
$(ad)_x^i$	i_x
$(ad)_x^r$	r_x

For the theoretical development of this and the succeeding chapter, the notation of Chapter 19 will be used.

Owing to the variations which take place between different schemes, it is not practicable to devise a set of symbols which shall have a well-recognized meaning for all schemes. The symbols used in pension fund work are usually limited to the service table symbols mentioned above, to those which are needed for representing the salary scale and functions closely related to it, and to commutation symbols in terms of which the values of the contributions and benefits may be expressed. The meanings of these symbols will be defined for each scheme in the light of the circumstances of that scheme. Typical symbols used in pension fund work will be introduced in this and the succeeding chapter. Although many of these symbols are widely used in Great Britain, they are not officially recognized, either in the International Notation or elsewhere, and their meanings may vary slightly according to the circumstances of the scheme in connexion with which they are used.

28·3. Salary scales

We have seen in Volume I that in valuing whole-life assurances and certain other classes of policy it is customary to group together all the policies on lives whose nearest age on the valuation date is x and to assume that all these lives are aged exactly x. A similar method is usually followed in pension fund valuations.*

Let us therefore imagine that there are N lives all aged exactly x. Let the total salaries per annum of these N lives be S. Then the number of lives expected to survive to age $x+t$ out of the N lives is $N(al)_{x+t}/(al)_x$. Since, however, the contributions and benefits in a pension scheme are normally functions of salary, we are often concerned more with the expected salaries of the survivors of the N lives in t years' time than with the actual number of survivors. In order to obtain the expected salaries we need a ratio of salary increase. Such a ratio is $\bar{s}_{x+t}/\bar{s}_x$, where $\bar{s}_x$ is defined by the statement that the expected salaries per annum, in t years' time, of the survivors of the N lives are

$$S\frac{\bar{s}_{x+t}}{\bar{s}_x}\cdot\frac{(al)_{x+t}}{(al)_x}. \qquad (28·1)$$

* Other methods are of course available. For example, when records are maintained on punched cards there is a gain in simplicity if the valuation age is taken as being $y+\frac{1}{2}$, where y is the difference between the calendar year of valuation and the calendar year of birth. The use of a method such as this will necessitate certain modifications to the valuation formulae. These modifications should present no difficulty and they will not be discussed in this book.

In practical pension fund work little use is found for the function $\bar{s}_x$, although it plays a part in the theoretical development of the subject. For reasons which will shortly be apparent, the function which is normally used in practical work in order to give effect to salary increases is s_x. This may be defined by the formula

$$s_x = \int_0^1 \bar{s}_{x+k}\, dk,^* \qquad (28\cdot2)$$

where $\bar{s}_x$ is defined as above.

An alternative way of defining s_x is to say that s_{x+t}/s_x is the ratio which the salary expected to be paid between ages $x+t$ and $x+t+1$ bears to the salary expected to be paid between ages x and $x+1$. While it is sometimes helpful to regard s_x in this way, the definition is not entirely satisfactory, since it does not indicate what allowance, if any, is to be made for the probability that the life or lives aged x will survive to receive the salary.

It will be clear from what has been said at the beginning of this section that in the practical application of formula (28·1) x is usually an integer. The formula, however, is equally applicable whether x is integral or fractional; it is also equally applicable whether t is integral or fractional. It is sometimes convenient to adopt the convention that both x and t are integers and to use the letter k to denote a fraction less than 1. When this is done, we require an expression for the expected salaries, in $t+k$ years' time, of the N persons now aged x. Such an expression is

$$S\frac{\bar{s}_{x+t+k}}{\bar{s}_x}\cdot\frac{(al)_{x+t+k}}{(al)_x}. \qquad (28\cdot3)$$

The salaries of many clerical staffs are reconsidered yearly on

* It would make for consistency if the definitions of $\bar{s}_x$ and s_x could be interchanged so that we could write

$$\bar{s}_x = \int_0^1 s_{x+k}\, dk.$$

This relationship would be in line with others to be given in this and the next chapter, such as

$$\bar{D}_x = \int_0^1 D_{x+k}\, dk$$

and

$$\bar{M}_x^{ia} = \int_0^1 M_{x+k}^{ia}\, dk.$$

However, the use of the symbol s_x in the sense in which it is used here is a long-accepted practice.

1 January; occasional special increases may be granted on other dates (for example on promotion to a more responsible position). Under these conditions it is usual in pension fund work to assume that all increases take place on 1 January. The effect of this assumption is that, for any one value of x (x being an integer and k a fraction less than 1), $\bar{s}_{x+k}$ is equal to s_x for all values of k. If, therefore, we are considering a group of employees assumed to be aged exactly x on the valuation date and if S is the number of salary units per annum which they will be receiving on the day after the valuation date, the expected number of salary units per annum which the survivors will be receiving at age $x+t+k$ (t being an integer) is

$$S\frac{s_{x+t}}{s_x}\cdot\frac{(al)_{x+t+k}}{(al)_x}. \qquad (28\cdot4)$$

It should be noted that formula (28·3) applies whatever may be the conditions under which increases of salary are granted; formula (28·4) is a special case applying only when salaries are revised yearly on 1 January.

A table of numerical values of s_x for integral values of x is referred to as the 'salary scale'. As a first step towards the construction of such a scale in respect of a clerical staff whose salaries are revised yearly on 1 January, it is usual to calculate, for various values of x, the average salary per annum actually being paid, as from the day after the valuation date, to members aged x. (The salaries coming into force on the day after the valuation date will normally be included in the valuation data.) Although the salary scale is not obtained by merely graduating these figures, consideration is given to them when the scale is being fixed. It is therefore sometimes convenient to adopt as a scale a series of figures which are of the same order of magnitude as these average salaries. In these circumstances s_x is usually referred to as the 'scale salary' at age x. It will be appreciated, however, that, as in the case of l_x, the absolute magnitude of s_x is really of no consequence; what matters is the ratio which s_{x+t} bears to s_x for any given values of x and t. Sometimes the scale is so constructed that $s_\alpha=1$, where α is the youngest age at which members are admitted to the fund. When this is done, s_x may be referred to as the 'salary scale ratio'.

Some employers of clerical staffs, instead of revising the salaries of all the employees yearly on a fixed date such as 1 January, reconsider each employee's salary yearly on his birthday or on the anniversary of his date of entry. A similar arrangement frequently applies to young workpeople. Older workpeople's salaries* are not normally revised except on promotion or on transfer to a different class of work, either of which events can take place at any time of the year. Under any of these conditions it may be assumed, for a large group of employees, that the salaries are revised continuously and therefore that $\bar{s}_x$ varies continuously with x.

In order to obtain a salary scale in respect of employees whose salaries are assumed to be revised continuously, it is usual to construct first a table of $\bar{s}_x$ for integral values of x, in order that consideration may be given to the average salaries actually being paid on the valuation date, and to obtain s_x from it by the approximation

$$s_x = \tfrac{1}{2}(\bar{s}_x + \bar{s}_{x+1}). \tag{28·5}$$

Occasionally it is necessary to value, as on 31 December, a fund the members of which are clerical staff whose salaries are revised yearly on some fixed date other than 1 January. (Also salaries may be revised yearly on 1 January and a valuation may be required on some date other than 31 December.) Under these conditions, let f be the period which has elapsed, on the valuation date, since the last yearly revision. Then, for any one value of x (x being an integer and k a fraction less than 1) $\bar{s}_{x+k}$ is equal to s_{x-f} for $0 \leqslant k < 1-f$ and to s_{x+1-f} for $1-f \leqslant k < 1$. It follows that (x being an integer)

$$s_x = (1-f)\, s_{x-f} + f s_{x+1-f}. \tag{28·6}$$

It may be convenient to construct a table of s_{x-f}, giving consideration to the average salaries actually being paid on the valuation date, and to obtain s_x from it by this relationship.

It would be possible to construct a multiple-decrement table based on salary units instead of on lives. In such a table we could have ${}^s(al)_x$ ($=\bar{s}_x(al)_x$), the number of salary units expected to be in force

* It is customary to refer to the pay of workpeople as wages rather than salary, but for convenience the term 'salary' is used in this book for both categories of employee.

at age x; ${}^s(ad)^w_x$, the number of salary units expected to withdraw between ages x and $x+1$; and so on. While the figures in the ${}^s(al)_x$ column of such a table would be quite useful for valuing the contributions, the figures in the other columns would be of very little use for valuing the benefits. Accordingly it is not customary to construct such a table. The normal procedure in pension fund work is to construct a service table, on the lines already described, and a table of monetary functions based on the service table, and to combine these monetary functions with s_x in order to obtain the necessary commutation functions for the valuation. When benefits on withdrawal, death and early retirement are valued, this procedure necessarily implies the assumption that the persons who withdraw, die and retire at any age x are random as regards their salaries. This assumption may not be in accordance with the facts since, for example, the average salaries of those who withdraw at age x may well tend to be less than the average salaries of those who remain in the service at the same age. However, the assumption is always implicitly made.

28·4. Future contributions (salaries revised on 1 January)

It is proposed in this section to consider the valuation of the employees' future contributions when salaries are revised yearly on 1 January. For this purpose it is convenient to introduce certain commutation functions. In the following symbols x is an integer (not necessarily the age on the valuation date) and k is a fraction less than 1:

$$D_x = v^x(al)_x \quad (\text{and } D_{x+k} = v^{x+k}(al)_{x+k}),$$

$$\bar{D}_x = \int_0^1 D_{x+k}\,dk \doteq \tfrac{1}{2}(D_x + D_{x+1}) \quad (\text{but } \bar{D}_{65} = 0),$$

$${}^sD_x = s_x D_x \quad (\text{but } {}^sD_{x+k} = s_x D_{x+k}),$$

$${}^s\bar{D}_x = \int_0^1 {}^sD_{x+k}\,dk = s_x \bar{D}_x,$$

$${}^s\bar{N}_x = \sum_{t=0}^{64-x} {}^s\bar{D}_{x+t}.$$

If s_x is large (say between 100 and 1,000) it is usual to take ${}^sD_x$ as being

$$\frac{s_x}{100}D_x,$$

in order to limit the number of figures in ${}^sD_x$. Throughout the remainder of this chapter and the next chapter it will be assumed that s_x is small (say between 1 and 10), so that when it is multiplied by commutation functions, such as D_x, division by 100 is not necessary.

Let us now consider a group of members aged x nearest birthday. Let the salaries per annum coming into force on the day after the valuation date be S_1, S_2, S_3, ... per annum and let the future contributions payable be r_1, r_2, r_3, ... per cent of these salaries. The r's may all be equal or they may vary with the age at entry. From formula (28·4), the value of the contributions payable by the members during the year of age $x+t$ to $x+t+1$ (t an integer) will be obtained by multiplying $\cdot 01\Sigma rS$ by

$$\int_0^1 v^{t+k}\frac{s_{x+t}}{s_x}\cdot\frac{(al)_{x+t+k}}{(al)_x}\,dk=\int_0^1\frac{{}^sD_{x+t+k}}{{}^sD_x}\,dk$$

$$=\frac{{}^s\bar{D}_{x+t}}{{}^sD_x}.$$

Therefore the factor by which $\cdot 01\Sigma rS$ must be multiplied in order to obtain the value of all the future contributions by the members is

$$\sum_{t=0}^{64-x}\frac{{}^s\bar{D}_{x+t}}{{}^sD_x}$$

$$=\frac{{}^s\bar{N}_x}{{}^sD_x}. \tag{28·7}$$

If, for some special reason, it is impracticable to include in the valuation data the salaries coming into force on the day after the valuation date, so that the salaries S_1, S_2, S_3, ... on which the valuation is based are the salaries per annum *on* the valuation date, it will be necessary either to multiply $\cdot 01\Sigma rS$ by s_x/s_{x-1} before applying the factor or to alter the factor to

$$\frac{{}^s\bar{N}_x}{s_{x-1}D_x}. \tag{28·8}$$

In order that the method of building up the functions may be perfectly clear, the headings of the columns for calculating numerical values of ${}^s\bar{N}_x/{}^sD_x$ are set out below.

x	v^x	$(al)_x$	D_x (2)×(3)	$\bar{D}_x$ $\frac{1}{2}(D_x+D_{x+1})$	s_x	${}^sD_x$ (4)×(6)	${}^s\bar{D}_x$ (5)×(6)	${}^s\bar{N}_x$ Σ(8)	${}^s\bar{N}_x/{}^sD_x$ (9)÷(7)
(1)	(2)	(3)	(4)	(5)	(6)	(7)	(8)	(9)	(10)

28·5. Future contributions (salaries revised continuously)

Let us now consider a group of members aged x nearest birthday whose conditions of employment are such that it may be assumed that the salaries are revised continuously. For this purpose it is convenient to introduce another commutation function:

$$^{\bar{s}}D_x = \bar{s}_x D_x \quad (\text{and } {}^{\bar{s}}D_{x+k} = \bar{s}_{x+k} D_{x+k}).$$

Let the salaries on the valuation date be $S_1, S_2, S_3, \ldots$ per annum and let the future contributions be $r_1, r_2, r_3, \ldots$ per cent of these salaries. From formula (28·3), the value of the contributions payable by the members during the year of age $x+t$ to $x+t+1$ (t an integer) will be obtained by multiplying $\cdot 01\Sigma rS$ by

$$\int_0^1 v^{t+k}\frac{\bar{s}_{x+t+k}}{\bar{s}_x}\cdot\frac{(al)_{x+t+k}}{(al)_x}dk = \int_0^1 \frac{{}^{\bar{s}}D_{x+t+k}}{{}^{\bar{s}}D_x}dk.$$

If we now introduce the approximation (x being any integer)

$$\begin{aligned}\int_0^1 {}^{\bar{s}}D_{x+k}\,dk &\doteq \int_0^1 \bar{s}_{x+k}\,dk \times \int_0^1 D_{x+k}\,dk\\ &= s_x\bar{D}_x\\ &= {}^s\bar{D}_x,\end{aligned}$$

this expression becomes $\dfrac{{}^s\bar{D}_{x+t}}{{}^{\bar{s}}D_x}.$

Therefore the factor for obtaining the value of all the future contributions by the members is

$$\sum_{t=0}^{64-x}\frac{{}^s\bar{D}_{x+t}}{{}^{\bar{s}}D_x}$$

$$=\frac{{}^s\bar{N}_x}{{}^{\bar{s}}D_x}. \qquad (28\cdot 9)$$

Comparing this formula with formula (28·7) we see that the numerators of the two formulae are identical and that whereas the denominator of formula (28·7) is ${}^sD_x = s_x D_x$ the denominator of formula (28·9) is ${}^{\bar{s}}D_x = \bar{s}_x D_x$. This is what we should expect, since the salaries S_1, S_2, S_3, ... to which formula (28·7) is applied are the salaries which come into force immediately after the granting of a yearly increase at age x, whereas the salaries to which formula (28·9) is applied are assumed to be revised continuously.

The headings of the columns required for evaluating the numerator of formula (28·9) will be exactly the same as the headings of columns (1) to (6), (8) and (9) in the table at the end of section 28·4. It will, however, be necessary to insert two further columns, $\bar{s}_x$ and ${}^{\bar{s}}D_x$, in order to evaluate the denominator. The final column in the table will be headed ${}^s\bar{N}_x/{}^{\bar{s}}D_x$.

28·6. Future contributions (salaries revised yearly but not on 1 January)

If salaries are revised yearly on some fixed date other than 1 January (or if they are revised yearly on 1 January and a valuation is required on some date other than 31 December), let f be the period which has elapsed, on the valuation date, since the last yearly revision.

Although $\bar{s}_x$ is not varying continuously, the procedure by which formula (28·9) was obtained will be applicable and the factor for valuing the contributions of a group of members aged x nearest birthday on the valuation date will be

$$\frac{{}^s\bar{N}_x}{{}^{\bar{s}}D_x}.$$

Since $\bar{s}_x$, when x is an integer, is equal to s_{x-f}, this becomes

$$\frac{{}^s\bar{N}_x}{s_{x-f}D_x}. \tag{28·10}$$

It will be seen that formulae (28·7) and (28·8) may be obtained from formula (28·10) by putting $f=0$ and $f=1$, respectively.

28·7. Future contributions (salaries revised continuously—continuous method)

If salaries are assumed to be revised continuously, an alternative method of valuing the contributions is available. This may be referred to as the continuous method in order to distinguish it from the method of section 28·5, which may be referred to as the yearly method.

With a continuous approach there is no need to represent time by two variables, t (an integer) and k (a fraction less than 1); it is sufficient to use one variable, t.

We shall need two new commutation functions, in addition to D_x and ${}^{s}D_x$:

$${}^{s}\bar{D}_x = \int_0^1 {}^{s}D_{x+t}\,dt \doteq \tfrac{1}{2}({}^{s}D_x + {}^{s}D_{x+1}) \quad (\text{but } {}^{s}\bar{D}_{65} = 0)$$

and
$${}^{s}\bar{N}_x = \int_0^{65-x} {}^{s}D_{x+t}\,dt = \sum_{t=0}^{64-x} {}^{s}\bar{D}_{x+t}.$$

From formula (28·1), the value of all the future contributions payable by a group of members aged x nearest birthday may be obtained by multiplying $\cdot 01\Sigma rS$ by

$$\int_0^{65-x} v^t \frac{\bar{s}_{x+t}}{\bar{s}_x} \cdot \frac{(al)_{x+t}}{(al)_x}\,dt = \int_0^{65-x} \frac{{}^{s}D_{x+t}}{{}^{s}D_x}\,dt$$
$$= \frac{{}^{s}\bar{N}_x}{{}^{s}D_x}. \qquad (28\cdot 11)$$

The order in which the functions are combined to build up the numerical value of this expression will differ slightly from that given in section 28·5 and is set out below.

x	v^x	$(al)_x$	D_x (2)×(3)	$\bar{s}_x$	${}^{s}D_x$ (4)×(5)	${}^{s}\bar{D}_x$ $\frac{1}{2}({}^{s}D_x+{}^{s}D_{x+1})$	${}^{s}\bar{N}_x$ Σ(7)	${}^{s}\bar{N}_x/{}^{s}D_x$ (8)÷(6)
(1)	(2)	(3)	(4)	(5)	(6)	(7)	(8)	(9)

Comparing this table with the procedure necessary for evaluating formula (28·9), we see that, whereas the table for formula (28·9) consists of 11 columns, that for formula (28·11) contains only nine. However, the formula that is generally used in practice is (28·9).

This is largely due to the fact that, in the early literature on pension funds, a yearly rather than a continuous approach was adopted; the resulting formulae have been generally accepted by the actuarial profession and are widely used. Also, in spite of the additional arithmetic involved, there is a considerable advantage in having one standard procedure for evaluating the numerator of the factor, regardless of the frequency of the salary revisions.

28·8. Contributions not varying directly with salary

Sometimes the contributions do not vary directly with the salary. For example, they may be a percentage of the amount by which the salary exceeds a fixed sum such as £100. In this event there are two alternative methods of procedure. One is to value the percentage of the total salary in the manner already described and to deduct from the value

$$(£100 \times \cdot 01r \times n_x)\frac{\overline{N}_x}{D_x},$$

where n_x is the number of members aged x nearest birthday and

$$\overline{N}_x = \sum_{t=0}^{64-x} \overline{D}_{x+t}.$$

Another method is to use a special salary scale which represents the amount by which the salary exceeds £100, so that the application of salary scale ratios, such as s_{x+t}/s_x, to the amount by which the total salaries at age x exceed £100 per member gives the expected amount by which the salaries will exceed £100 per member at a future date.

Other variations do not lend themselves to such simple treatment. For example, it is by no means uncommon for the rules of a pension fund to contain a provision that all salaries over a certain figure shall be treated as being equal to that figure for the purpose of calculating contributions and benefits. Unless the limiting salary is a high one, so that few members are likely to reach it, this provision is a troublesome one to deal with, since the various members in any one age group may have widely differing salaries on the valuation date and may therefore be expected to reach the limiting salary at different ages.

In these circumstances, an accurate valuation is not possible without a prohibitive amount of labour and it may be necessary to use approximate methods, the results of which will have to be interpreted with the skill and judgment that are developed only with training and experience. In this book it is proposed to deal only with comparatively simple problems which are susceptible to treatment by straightforward methods. We may, however, consider the value of the future contributions of a member aged x whose salary is S and whose rate of contribution is r per cent, in a fund where the maximum salary for the purpose of contributions and benefits is £1,000.

Let n be an integer such that

$$S\frac{s_{x+n-1}}{s_x} \leqslant 1{,}000 < S\frac{s_{x+n}}{s_x}.$$

Then the value of the contributions may be taken as

$$\cdot 01r\left\{S\frac{{}^s\overline{N}_x - {}^s\overline{N}_{x+n}}{{}^sD_x} + 1{,}000\frac{\overline{N}_{x+n}}{D_x}\right\}. \qquad (28\cdot 12)$$

Although this formula has the appearance of accuracy, it is not entirely satisfactory, since it assumes that the salary of the individual member will progress exactly according to the scale; any deviation from the scale at a future date may affect not only the contribution payable at that date but also the age at which the figure of £1,000 is reached.

28·9. Employer's contributions

If the employer's contributions are a percentage of salaries, they may be valued in the same way as employees' contributions. In some funds the employer contributes at the end of each year an amount which is equal to that which has been contributed by the employees during the year. Under these conditions the value of the employer's contributions may be obtained by multiplying the value of the employees' contributions by $v^{\frac{1}{2}}$.

Often when a fund is set up there is an outstanding liability for past service pensions. This liability is usually liquidated over a series of years by special contributions from the employer. These contributions may be valued by an annuity-certain.

EXAMPLES

Example 28·1

On the basis of the rates of death, retirement and withdrawal and the salary scale reproduced on pages 246 and 248 of this book and assuming 3 per cent interest, calculate the value of the future contributions for the following portion of a tabulation of members of a pension scheme. Assume that the valuation is made on 31 December 1956 and that

(1) salaries are revised on 1 January each year,

(2) salaries are revised on 1 April each year,

(3) salaries are revised on the anniversary of the employee's entry into service.

Age on valuation date	Age at entry	Contribution rate (%)	Salary on 1 January 1957
49	28	4	1,000
	32	4·5	600
	37	5	800
50	28	4	1,500
	32	4·5	700
	37	5	700

Solution

The annual contributions on the basis of the salaries in force on 1 January 1957 are

Valuation age	Contributions
49	107
50	126·5

If the salaries are revised on 1 January each year the valuation can be made by multiplying these contributions by ${}^s\overline{N}_x/{}^sD_x$.

From the tables we have

x	D_x	s_x	${}^sD_x$ (2)×(3)	${}^s\overline{N}_x$	${}^s\overline{N}_x/{}^sD_x$ (5)÷(4)	Contributions	Value (6)×(7)
(1)	(2)	(3)	(4)	(5)	(6)	(7)	(8)
49	17,966	3·204	57,563	647,771	11·25	107	1204
50	17,277	3·250	56,150	591,313	10·53	126·5	1332
					Total value		2536

If the salaries are revised on 1 April each year, the valuation can be made by multiplying the contributions by ${}^s\overline{N}_x/s_{x-\frac{3}{4}}D_x$. Hence, we have

x	s_{x-1}	s_x	$s_{x-\frac{3}{4}}$	${}^s\overline{N}_x/s_{x-\frac{3}{4}}D_x$	Value
49	3·156	3·204	3·168	11·38	1218
50	3·204	3·250	3·216	10·64	1346
				Total value	2564

If the salaries are revised on the anniversary of the employee's entry into service, we may assume that the anniversaries are distributed uniformly over the year and use the formula for salaries revised continuously. This is ${}^s\overline{N}_x/{}^{\bar{s}}D_x$. Hence, we have

x	s_{x-1}	s_x	$\bar{s}_x$	${}^s\overline{N}_x/{}^{\bar{s}}D_x$	Value
49	3·156	3·204	3·180	11·34	1213
50	3·204	3·250	3·227	10·61	1342
				Total value	2555

Example 28·2

A pension scheme provides that all salary increases after age 50 shall be ignored for the purpose of determining contributions and benefits. Assuming the rates of death, retirement and withdrawal and the salary scale reproduced on pages 246 and 248 of this book and 3 per cent interest, calculate the factor for valuing future contributions as on 31 December for a group of lives aged 30. Salaries are revised yearly on 1 January.

Solution

For contributions up to age 50, the factor to be applied to $\cdot 01\Sigma rS$ is

$$\frac{{}^s\overline{N}_{30}-{}^s\overline{N}_{50}}{{}^sD_{30}}.$$

For contributions after age 50, the factor is

$$\frac{{}^sD_{50}}{{}^sD_{30}}\cdot\frac{\overline{N}_{50}}{D_{50}}.$$

Hence, the combined factor is

$$\frac{{}^s\bar{N}_{30} - {}^s\bar{N}_{50} + s_{50}\bar{N}_{50}}{s_{30}D_{30}} = \frac{1,876,233 - 591,313 + 3\cdot250 \times 171,308}{1\cdot950 \times 34,250}$$

$$= \frac{1,876,233 - 591,313 + 556,751}{66,787}$$

$$= \frac{1,841,671}{66,787} = 27\cdot58.$$

Rough check

The answer is slightly less than ${}^s\bar{N}_{30}/{}^sD_{30}$, as it should be.

CHAPTER 29

PENSION FUNDS—VALUATION OF BENEFITS

29·1. Introduction

The benefits provided by pension funds can be divided into two main classes; pensions payable on retirement and cash payments on death and withdrawal.

Pensions usually depend on the average salary throughout service, the average salary during the last few years of service or the salary at the date of retirement (referred to as the final salary). The pension usually consists of $1/m$th of this average or final salary for each year of service; m is a number, such as 60, which is fixed for all members. Complications may be introduced such as a limit to the number of years' service which may count for pension and the counting of complete years' service only.

It is proposed in this chapter to consider first the valuation of various types of pension, then the valuation of cash payments on death and withdrawal and finally the calculation of rates of contribution for new entrants.

In sections 4, 5 and 6 of the preceding chapter the problem of valuing the contributions was approached by a yearly method on the basis of various assumptions as to salary revisions; it was found that the numerator of the factor is a function of s_x which is independent of the conditions under which salaries are revised, and that the denominator is ${}^sD_x$ if salaries are revised yearly on 1 January, ${}^{\bar{s}}D_x$ if they are revised continuously and $s_{x-f}D_x$ if they are revised yearly and the last revision took place at age $x-f$. In section 7 of the same chapter a continuous approach was made on the assumption of continuous salary revisions and this led to a factor of which the numerator is a function of $\bar{s}_x$ and the denominator is ${}^{\bar{s}}D_x$. Similar principles apply to the factors for valuing the benefits; this fact will be illustrated in sections 2 and 3 of this chapter where the valuation of one type of pension will be considered on the basis of all three assumptions as to salary revisions. In subsequent sections of the chapter it will be assumed that salaries are revised yearly on 1 January and the variations in the factors consequent upon the other two assumptions will not be investigated.

29·2. Ill-health pensions based on average salary

Let us first consider the value of a pension, payable on ill-health retirement, of $1/m$th of the average salary received throughout service for each year of service, fractions of a year counting. This pension has to be valued in two parts:

(i) the pension which has already accrued to each member in respect of the salary which he has received before the valuation date (this is known as the past-service pension),

(ii) the pension which will accrue to each member in respect of salary payments which he will receive after the valuation date (this is known as the future-service pension).

Since the factor for valuing the past-service pension is independent of the salary scale, there is no point in adopting a yearly approach. We shall need two commutation functions (in addition to D_x which was defined in Chapter 28):

$$C_x^{ia} = \int_0^1 D_{x+t}(a\mu)_{x+t}^i \bar{a}_{x+t}^i \, dt \doteq v^{x+\frac{1}{2}}(ad)_x^i \bar{a}_{x+\frac{1}{2}}^i$$

and

$$M_x^{ia} = \int_0^{65-x} D_{x+t}(a\mu)_{x+t}^i \bar{a}_{x+t}^i \, dt = \sum_{t=0}^{64-x} C_{x+t}^{ia},$$

where $\bar{a}_{x+t}^i$ is the value of an annuity of 1 to a pensioner aged $x+t$ who has just retired owing to ill-health.

The past-service pension of each member is $1/m$th of his total past salary. The value of the past-service ill-health pensions of a group of members aged x nearest birthday is obtained by multiplying $1/m$th of their total past salaries by

$$\int_0^{65-x} v^t \frac{(al)_{x+t}}{(al)_x} (a\mu)_{x+t}^i \bar{a}_{x+t}^i \, dt$$

$$= \frac{M_x^{ia}}{D_x}. \tag{29·1}$$

When the pension is based on the average salary throughout service, the valuation data will usually include the total past salary, or alternatively the past-service pension, of each member. If this is not the case, an approximation to the total past salary of a member aged x nearest birthday who entered at y nearest birthday may be obtained by multiplying his salary coming into force on the day after the valuation date by the ratio

$$(s_y + s_{y+1} + \ldots + s_{x-1})/s_x.$$

(It is assumed here that the salaries are revised yearly on 1 January and that the salaries coming into force on the day after the valuation date are included in the valuation data. When these assumptions are not fulfilled, the denominator of the ratio will be suitably altered.) The use of the salary scale for estimating past salaries is, however, not normally to be recommended.

For the purpose of applying these ratios, the members may be grouped according to age (nearest birthday) attained and sub-grouped according to age (nearest birthday) at entry. Alternatively they may be grouped according to age attained and the ratio for each group may be based on the average age at entry for the group.

For valuing the future-service pensions by the yearly method, three further commutation functions may be introduced:

$$\overline{M}^{ia}_x = \int_0^1 M^{ia}_{x+k}\,dk \doteq \tfrac{1}{2}(M^{ia}_x + M^{ia}_{x+1}) = M^{ia}_x - \tfrac{1}{2}C^{ia}_x = M^{ia}_{x+1} + \tfrac{1}{2}C^{ia}_x,$$

$$^s\overline{M}^{ia}_x = s_x \overline{M}^{ia}_x$$

and

$$^s\overline{R}^{ia}_x = \sum_{t=0}^{64-x} {}^s\overline{M}^{ia}_{x+t}.$$

Let the total salaries per annum, coming into force on the day after the valuation date, of a group of members aged x nearest birthday (assumed to be aged exactly x) be ΣS. Each future salary payment may be regarded as carrying the right to a pension of $1/m$th of its amount, to be entered upon on retirement. The value of the ill-health pensions secured by the salary payments which the members will receive during the year of age $x+t$ to $x+t+1$ (t an integer) is therefore obtained by multiplying $1/m$th of ΣS by

$$\int_0^1 v^{t+k}\frac{s_{x+t}}{s_x}\cdot\frac{(al)_{x+t+k}}{(al)_x}\cdot\frac{M^{ia}_{x+t+k}}{D_{x+t+k}}\,dk = \int_0^1 \frac{s_{x+t}M^{ia}_{x+t+k}}{^sD_x}\,dk$$

$$= \frac{s_{x+t}\overline{M}^{ia}_{x+t}}{^sD_x}$$

$$= \frac{^s\overline{M}^{ia}_{x+t}}{^sD_x}.$$

Therefore the factor by which $1/m$th of ΣS must be multiplied in order to obtain the value of all the future-service ill-health pensions for the group is

$$\sum_{t=0}^{64-x} \frac{{}^s\overline{M}^{ia}_{x+t}}{{}^sD_x}$$

$$=\frac{{}^s\overline{R}^{ia}_x}{{}^sD_x}. \qquad (29{\cdot}2)$$

If salaries are assumed to be revised continuously and if a yearly method is adopted, the factor in respect of the salary received during the year of age $x+t$ to $x+t+1$ is

$$\int_0^1 v^{t+k}\frac{\bar{s}_{x+t+k}}{\bar{s}_x}\cdot\frac{(al)_{x+t+k}}{(al)_x}\cdot\frac{M^{ia}_{x+t+k}}{D_{x+t+k}}\,dk=\int_0^1\frac{\bar{s}_{x+t+k}M^{ia}_{x+t+k}}{{}^{\bar{s}}D_x}\,dk.$$

If we now introduce the approximation (x being any integer)

$$\int_0^1 \bar{s}_{x+k}M^{ia}_{x+k}\,dk \doteqdot \int_0^1 \bar{s}_{x+k}\,dk\times\int_0^1 M^{ia}_{x+k}\,dk$$

$$=s_x\overline{M}^{ia}_x$$

$$={}^s\overline{M}^{ia}_x,$$

this expression becomes

$$\frac{{}^s\overline{M}^{ia}_{x+t}}{{}^{\bar{s}}D_x}$$

and the factor in respect of all the future salaries is

$$\sum_{t=0}^{64-x}\frac{{}^s\overline{M}^{ia}_{x+t}}{{}^{\bar{s}}D_x}$$

$$=\frac{{}^s\overline{R}^{ia}_x}{{}^{\bar{s}}D_x}. \qquad (29{\cdot}3)$$

As might have been expected, the numerator of this formula is identical with the numerator of formula (29·2) and the denominator is ${}^{\bar{s}}D_x$ instead of ${}^sD_x$.

If salaries are revised yearly at age $x-f$ (x an integer) the procedure by which formula (29·3) was obtained will be applicable and the factor will be

$$\frac{{}^s\overline{R}^{ia}_x}{{}^{\bar{s}}D_x}.$$

Since $\bar{s}_x$, when x is an integer, is equal to s_{x-f}, this becomes

$$\frac{{}^s\overline{R}^{ia}_x}{s_{x-f}D_x}. \qquad (29{\cdot}4)$$

The headings of the columns for calculating numerical values of ${}^s\bar{R}^{ia}_x$ will be as follows:

x	$v^{x+\frac{1}{2}}$	$(ad)^i_x$	$\bar{a}^i_{x+\frac{1}{2}}$	C^{ia}_x (2)×(3)×(4)	M^{ia}_x Σ(5)	$\bar{M}^{ia}_x$ $\frac{1}{2}(M^{ia}_x+M^{ia}_{x+1})$	s_x	${}^s\bar{M}^{ia}_x$ (7)×(8)	${}^s\bar{R}^{ia}_x$ Σ(9)
(1)	(2)	(3)	(4)	(5)	(6)	(7)	(8)	(9)	(10)

In most service tables $(ad)^i_x$ is equal to zero at the youngest ages (say under age 30). At these ages there will be no figures in columns (2), (3), (4) and (5). All the columns will end at age 64.

On the assumption that $\bar{M}^{ia}_x$ is exactly equal to $\frac{1}{2}(M^{ia}_x+M^{ia}_{x+1})$, it can easily be shown that the numerator of formula (29·2) is equal to

$$\tfrac{1}{2}s_x C^{ia}_x + (s_x+\tfrac{1}{2}s_{x+1})\,C^{ia}_{x+1} + (s_x+s_{x+1}+\tfrac{1}{2}s_{x+2})\,C^{ia}_{x+2} + \dots \quad (29\cdot5)$$

This is obviously approximately correct if we assume (as we do in effect when we use the approximation $C^{ia}_x \doteq v^{x+\frac{1}{2}}(ad)^i_x\bar{a}^i_{x+\frac{1}{2}}$) that the value of the pensions is the same as that which would be developed if all the retirements occurring in each year of age were concentrated at the middle of the year.

In pension fund work it is customary to use the symbols C and M, without a bar, in respect of lump sum benefits which are payable, and annuities which commence to be payable, immediately on death, withdrawal or retirement. This is inconsistent with the use of $\bar{C}_x$ and $\bar{M}_x$ in ordinary Life Contingencies. It is justified partly by the fact that in pension fund work we have no use for functions which provide for payment at the end of the year of death, withdrawal or retirement and partly by the convenience of being able to use relationships like

$$\bar{M}^{ia}_x = \int_0^1 M^{ia}_{x+k}\,dk \doteq \tfrac{1}{2}(M^{ia}_x+M^{ia}_{x+1})$$

by analogy with the relationship in ordinary Life Contingencies (see Volume I, Chapter 4)

$$\bar{D}_x = \int_0^1 D_{x+t}\,dt \doteq \tfrac{1}{2}(D_x+D_{x+1}).$$

29·3. Future-service ill-health pensions based on average salary—continuous method

In the preceding section the past-service pensions were valued by the continuous method and the future-service pensions by the yearly method. If salaries are assumed to be revised continuously, the future-service pensions may be valued by the continuous method.

For this purpose we may use three more commutation functions:

$$^{\bar{s}}M_x^{ia} = \bar{s}_x M_x^{ia},$$

$$^{\bar{s}}\overline{M}_x^{ia} = \int_0^1 {}^{\bar{s}}M_{x+t}^{ia}\,dt \doteq \tfrac{1}{2}({}^{\bar{s}}M_x^{ia} + {}^{\bar{s}}M_{x+1}^{ia})$$

and
$$^{\bar{s}}\overline{R}_x^{ia} = \int_0^{65-x} {}^{\bar{s}}M_{x+t}^{ia}\,dt = \sum_{t=0}^{64-x} {}^{\bar{s}}\overline{M}_{x+t}^{ia}.$$

The value of the future-service ill-health pensions may be obtained by multiplying 1/*m*th of ΣS by

$$\int_0^{65-x} v^t \frac{\bar{s}_{x+t}}{\bar{s}_x} \cdot \frac{(al)_{x+t}}{(al)_x} \cdot \frac{M_{x+t}^{ia}}{D_{x+t}}\,dt = \int_0^{65-x} \frac{^{\bar{s}}M_{x+t}^{ia}}{^{\bar{s}}D_x}\,dt$$

$$= \frac{^{\bar{s}}\overline{R}_x^{ia}}{^{\bar{s}}D_x}. \qquad (29\cdot6)$$

For calculating numerical values of $^{\bar{s}}\overline{R}_x^{ia}$, ten columns will be needed. The headings of the first six columns will be the same as for $^{s}\overline{R}_x^{ia}$; the headings of the other four columns will be as follows:

$\bar{s}_x$	$^{\bar{s}}M_x^{ia}$ (6) × (7)	$^{\bar{s}}\overline{M}_x^{ia}$ $\tfrac{1}{2}({}^{\bar{s}}M_x^{ia} + {}^{\bar{s}}M_{x+1}^{ia})$	$^{\bar{s}}\overline{R}_x^{ia}$ $\Sigma(9)$
(7)	(8)	(9)	(10)

On the assumption that $^{\bar{s}}\overline{M}_x^{ia}$ is exactly equal to $\tfrac{1}{2}({}^{\bar{s}}M_x^{ia} + {}^{\bar{s}}M_{x+1}^{ia})$, it can easily be shown that

$$^{\bar{s}}\overline{R}_x^{ia} = \tfrac{1}{2}\bar{s}_x C_x^{ia} + (\tfrac{1}{2}\bar{s}_x + \bar{s}_{x+1})\,C_{x+1}^{ia} + (\tfrac{1}{2}\bar{s}_x + \bar{s}_{x+1} + \bar{s}_{x+2})\,C_{x+2}^{ia} + \ldots. \qquad (29\cdot7)$$

This can be verified in the same way as formula (29·5).

For the reasons explained in section 28·7, formula (29·6) is not generally used in practice.

29·4. Age pensions based on average salary

The procedure set out in section 2 has to be modified slightly for age retirements because it is necessary to allow not only for those who retire before age 65 but also for those who retire immediately on attaining that age. We therefore define C_x^{ra} in the same way as C_x^{ia} for all ages up to 64 inclusive but for age 65 we write

$$C_{65}^{ra} = D_{65}\,\bar{a}_{65}^{r}.$$

We also write
$$M_x^{ra} = \sum_{t=0}^{65-x} C_{x+t}^{ra},$$

so that the value of C_{65}^{ra} is included in the summation. The functions $\bar{M}_x^{ra}$, ${}^s\bar{M}_x^{ra}$ and ${}^s\bar{R}_x^{ra}$ are defined, for all ages up to 64 inclusive, in the same way as the corresponding functions for ill-health retirement; when $x=65$, all these functions are zero.

The factor for valuing the past-service age pensions is therefore

$$\frac{M_x^{ra}}{D_x} \tag{29·8}$$

and the factor for valuing the future-service age pensions is

$$\frac{{}^s\bar{R}_x^{ra}}{{}^sD_x}. \tag{29·9}$$

We have already seen that when a table of factors for valuing ill-health pensions is being constructed, all the columns end at age 64. The summations therefore begin at that age (because the columns are summed from the bottom upwards). When we are dealing with age retirement, C_x^{ra} has to be summed from age 65 in order to obtain M_x^{ra}, but ${}^s\bar{M}_x^{ra}$ is zero at age 65 and is therefore summed from age 64 only. This process automatically provides the correct future-service pension ($1/m$th of the salary received between age x and age 65) for a member who is aged x on the valuation date and who retires at exact age 65.

It will be noted that the approximation

$$C_{60}^{ra} \doteq v^{60\frac{1}{2}}(ad)_{60}^{r}\,\bar{a}_{60\frac{1}{2}}^{r}$$

is not very satisfactory if there is a concentration of age retirements at exact age 60. Although a more accurate expression could be developed the point is usually ignored in practical work.

If the age pension is the same as the ill-health pension, the factors for the two types of retirement may be combined and we may write, for example,

$$M_x^{ia}+M_x^{ra}=M_x^{(i+r)a}.$$

29·5. Ill-health pensions based on average salary—only completed years counting

The rules of many pension schemes provide that only completed years of service shall count for pension. When the pension is based on average salary, the effect of this provision is that each member receives no pension in respect of the salary which is paid to him during the year of service in which he retires.

With a provision of this kind, there is scope for a difference of opinion as to what constitutes 'past-service pension'. We may interpret the term as meaning either

(i) the pension to which the member would be entitled if he retired on the valuation date (i.e. $1/m$th of the total salary which he received up to the end of the last completed year of his service), or

(ii) $1/m$th of the total salary which the member has received up to the valuation date.

These two variants of the past-service pension may be referred to for convenience as 'curtate' and 'complete' respectively. In either case the future-service pension will be equal to the difference between the total pension and the past-service pension.

In the absence of legal requirements or other considerations, it is natural to choose the interpretation which conforms to the manner in which the valuation data have been prepared. That is to say, if the record which the employer keeps in respect of each member is brought up to date at the end of each year of service, the 'total past salary' supplied by the employer will relate to completed years only and the past-service pension will normally be regarded as being $1/m$th of this figure, i.e. it will be curtate; if the employer quotes, in respect of each member, the total past salary up to the valuation date, it will be natural to take the past-service pension as being $1/m$th of this, i.e. it will be complete.

Let us first assume that the past-service pensions are curtate.

The factor for valuing the past-service ill-health pensions will be

$$\frac{M_x^{ia}}{D_x},$$

which is formula (29·1).

For the future-service pensions, two further commutation functions are required:

$$^sM_x^{ia} = s_{x-1}M_x^{ia}$$

and
$$^sR_x^{ia} = \sum_{t=0}^{64-x} {^sM_{x+t}^{ia}}.$$

In order to approach the problem of valuing the future-service pensions, it is convenient to consider an individual member aged x whose salary, coming into force on the day after the valuation date, is S and whose past service is $n+f$ years (n being an integer and f a fraction less than 1). Since his past-service pension is curtate, his 'future' service commenced at age $x-f$. He will be entitled to a pension equal to $1/m$th of the salary which he receives during the year from age $x+t-f$ to age $x+t+1-f$ (where t can be any integer from 0 to $64-x$, both inclusive) if he retires at any time after age $x+t+1-f$. His expected salary for this year is

$$S\frac{\int_0^1 \bar{s}_{x+t-f+k}\,dk}{s_x} = S\frac{s_{x+t-f}}{s_x}.$$

The value of the ill-health pension secured by this year's salary is therefore obtained by multiplying $1/m$th of S by

$$v^{t+1-f}\frac{s_{x+t-f}}{s_x}\cdot\frac{(al)_{x+t+1-f}}{(al)_x}\cdot\frac{M_{x+t+1-f}^{ia}}{D_{x+t+1-f}} = \frac{^sM_{x+t+1-f}^{ia}}{^sD_x}.$$

Therefore the factor by which $1/m$th of S must be multiplied in order to obtain the value of the whole of the member's future-service ill-health pension is

$$\sum_{t=0}^{64-x}\frac{^sM_{x+t+1-f}^{ia}}{^sD_x}$$

$$= \frac{^sR_{x+1-f}^{ia}}{^sD_x}. \qquad (29·10)$$

For a group of members aged x nearest birthday, the anniversaries of whose entry dates are distributed at random over the calendar year, it may be assumed that the average value of f is $\frac{1}{2}$. Putting $f=\frac{1}{2}$ in formula (29·10), we have as the factor for valuing the future-service ill-health pensions of the group, when the past-service pensions are curtate,

$$\frac{{}^sR^{ia}_{x+\frac{1}{2}}}{{}^sD_x}. \tag{29·11}$$

For practical purposes, this may be taken as being equal to

$$\frac{\frac{1}{2}({}^sR^{ia}_x+{}^sR^{ia}_{x+1})}{{}^sD_x}=\frac{{}^sR^{ia}_x-\frac{1}{2}{}^sM^{ia}_x}{{}^sD_x}=\frac{{}^sR^{ia}_{x+1}+\frac{1}{2}{}^sM^{ia}_x}{{}^sD_x}. \tag{29·12}$$

Let us now assume that the past-service pensions are complete.

It is again convenient to consider an individual member whose past service is $n+f$ years. It will be appreciated that the effect of making the past-service pension complete instead of curtate is merely to transfer from future service to past service the pension to which the member will be entitled, on retirement after age $x+1-f$, in respect of the salary which he received between age $x-f$ and age x.

Strictly speaking, in order to obtain the value of the complete past-service pension, we should multiply $1/m$th of the salary which the member received before age $x-f$ by

$$\frac{M^{ia}_x}{D_x},$$

i.e. by formula (29·1), and we should multiply $1/m$th of the salary which he received between age $x-f$ and age x by

$$\frac{M^{ia}_{x+1-f}}{D_x}.$$

In practice, it is not possible to split the past-service pensions up in this way, nor would it be worth while even if it were possible. It is sufficient to obtain the value of the complete past-service ill-health pensions of a group of members by multiplying $1/m$th of their total past salaries by formula (29·1).

Coming now to the future-service pension, we have to consider by how much formula (29·10) must be reduced in order to allow for the fact that the service between ages $x-f$ and x is not now included in the future service. The salary received during this period (if the salary has progressed according to scale) was

$$Sf\frac{s_{x-1}}{s_x}$$

and, as explained above, if the pension in respect of this salary is valued as past-service pension, the factor used in practice is M_x^{ia}/D_x and not the theoretically accurate factor M_{x+1-f}^{ia}/D_x. The factor for valuing the future-service ill-health pensions, when the past-service pensions are complete, is therefore

$$\frac{{}^sR_{x+1-f}^{ia}}{{}^sD_x}-\frac{fs_{x-1}}{s_x}\cdot\frac{M_x^{ia}}{D_x}=\frac{{}^sR_{x+1-f}^{ia}-f{}^sM_x^{ia}}{{}^sD_x}$$

$$\doteq\frac{{}^sR_{x+1}^{ia}}{{}^sD_x}. \tag{29·13}$$

The numerator of formula (29·12) is equal to

$$\tfrac{1}{2}s_{x-1}C_x^{ia}+(\tfrac{1}{2}s_{x-1}+s_x)\,C_{x+1}^{ia}+(\tfrac{1}{2}s_{x-1}+s_x+s_{x+1})\,C_{x+2}^{ia}+\dots \tag{29·14}$$

and the numerator of formula (29·13) is equal to

$$s_xC_{x+1}^{ia}+(s_x+s_{x+1})\,C_{x+2}^{ia}+\dots. \tag{29·15}$$

In order to verify that formula (29·14) is approximately correct, we may consider a member aged x whose past service is $n+\frac{1}{2}$ years. For example, let us assume that the member retires during the year following the date on which he attains age $x+2$. If he retires in the first half of this year, he will be entitled to a future-service pension in respect of the salary which he received from $x-\frac{1}{2}$ to age $x+1\frac{1}{2}$; if he retires in the second half of the year, he will be entitled to a future-service pension in respect of the salary which he received from age $x-\frac{1}{2}$ to age $x+2\frac{1}{2}$. As a practical approximation, therefore, we may say that, on the average, he will be entitled on retirement during this year to a future-service pension in respect of the salary which he received from age $x-\frac{1}{2}$ to age $x+2$; in salary scale units this is represented by $\frac{1}{2}s_{x-1}+s_x+s_{x+1}$. This confirms the third term in formula (29·14); the other terms in the formula may be confirmed in the same way. Formula (29·15) differs from formula (29·14) by $\frac{1}{2}s_{x-1}M_x^{ia}$, which is as it should be.

29·6. Age pensions based on average salary—only completed years counting

So far as past-service pensions are concerned, the considerations set out in the preceding section in respect of ill-health retirements apply equally to age retirements. It follows that, whether the past-service pensions are curtate or complete, a suitable factor for valuing the past-service age pensions is

$$\frac{M_x^{ra}}{D_x},$$

which is formula (29·8).

Care is needed when valuing the future-service age pensions, because those who retire at exact age 65, like those who retire at earlier ages, will be entitled to no pension in respect of the salary which they receive during the year of service in which they retire. The method of the preceding section, when applied to age retirement, makes suitable provision for those who retire before age 65, provided that their dates of retirement are random in relation to the anniversaries of their dates of entry. In order to provide correctly for those who retire at exact age 65 we write

$${}^sM_x^{ra} = s_{x-1}M_x^{ra} \quad \text{when } x \leqslant 64,$$

$${}^sM_{65}^{ra} = \tfrac{1}{2}s_{64}M_{65}^{ra}$$

and

$${}^sR_x^{ra} = \sum_{t=0}^{65-x} {}^sM_{x+t}^{ra}.$$

When the past-service pensions are curtate, the factor for valuing the future-service age pensions will be

$$\frac{{}^sR_{x+\frac{1}{2}}^{ra}}{{}^sD_x}, \tag{29·16}$$

which may be taken as being equal to

$$\frac{\tfrac{1}{2}({}^sR_x^{ra} + {}^sR_{x+1}^{ra})}{{}^sD_x} = \frac{{}^sR_x^{ra} - \tfrac{1}{2}{}^sM_x^{ra}}{{}^sD_x} = \frac{{}^sR_{x+1}^{ra} + \tfrac{1}{2}{}^sM_x^{ra}}{{}^sD_x}, \tag{29·17}$$

except at age 65, where the factor is zero.

When the past-service pensions are complete, the factor for valuing the future-service age pensions will be

$$\frac{{}^{s}R^{ra}_{x+1}}{{}^{s}D_x}. \tag{29·18}$$

The foregoing procedure implies the assumption that the members who retire at exact age 65 will be entitled to pensions equal to $1/m$th of the salaries which they receive up to age $64\frac{1}{2}$. This is approximately correct on the average if the dates of entry are random in relation to the birthdays.

In some schemes it is a condition that all employees who enter the employer's service before attaining a certain age (say 25) shall be required to join the scheme immediately on attaining that age; salary received before that age does not rank for pension. Such members will be entitled, on retirement at age 65, to a pension based on the salaries which they receive up to age 65. In these circumstances ${}^{s}M^{ra}_{65}$ must be taken as being equal to $s_{64}M^{ra}_{65}$.

29·7. Pensions based on final salary

Let us now consider the value of a pension, payable on ill-health retirement, of $1/m$th of the salary at the date of retirement (the 'final salary') for each year of service.

The valuation data will usually include the exact date of entry of each member. In order to avoid fractions it is convenient to calculate each member's past service to the nearest integer, or alternatively to obtain it by subtracting the nearest age at entry from the nearest age on the valuation date, and to regard the resulting figure as being the member's exact past service.

For valuing the past-service ill-health pensions, we shall require two more commutation functions:

$${}^{z}C^{ia}_{x} = s_x C^{ia}_{x}$$

and

$${}^{z}M^{ia}_{x} = \sum_{t=0}^{64-x} {}^{z}C^{ia}_{x+t}.$$

Let n be the past service of an individual member aged x nearest birthday and let S be his salary coming into force on the day after the valuation date. Then the total past-service pensions of a group of members aged x nearest birthday will be $1/m$th of ΣnS and the value of the past-service ill-health pensions, with allowance for future increases of salary according to scale, will be obtained by multiplying $1/m$th of ΣnS by

$$\sum_{t=0}^{64-x} v^t \frac{s_{x+t}}{s_x} \cdot \frac{(al)_{x+t}}{(al)_x} \cdot \frac{C_{x+t}^{ia}}{D_{x+t}} = \sum_{t=0}^{64-x} \frac{{}^zC_{x+t}^{ia}}{{}^sD_x}$$

$$= \frac{{}^zM_x^{ia}}{{}^sD_x}. \qquad (29\cdot19)$$

It simplifies the approach to the problem of valuing the future-service ill-health pensions if we assume first that fractions of a year do not count. On this assumption, since the past service is regarded as being an integer, a member aged x will receive no pension in respect of his service between ages $x+t$ and $x+t+1$ (t an integer) unless he retires after attaining age $x+t+1$.

Let

$${}^zR_x^{ia} = \sum_{t=0}^{64-x} {}^zM_{x+t}^{ia}.$$

Then the value of the ill-health pensions to which a group of members aged x nearest birthday will be entitled in respect of their service between ages $x+t$ and $x+t+1$ is equal to $1/m$th of ΣS multiplied by

$$\frac{{}^zM_{x+t+1}^{ia}}{{}^sD_x}.$$

The factor for valuing the future-service ill-health pensions, when fractions of a year do not count, is therefore

$$\sum_{t=0}^{63-x} \frac{{}^zM_{x+t+1}^{ia}}{{}^sD_x}$$

$$= \frac{{}^zR_{x+1}^{ia}}{{}^sD_x}. \qquad (29\cdot20)$$

For valuing the future-service ill-health pensions when fractions of a year count, we need a definition of ${}^zM_{x+k}^{ia}$ (x an integer and k a fraction less than 1); the definition of ${}^zM_x^{ia}$ given above in terms of ${}^zC_x^{ia}$ applies only when x is an integer.

Let

$${}^zM_{x+k}^{ia} = {}^zM_{x+1}^{ia} + s_x \int_k^1 D_{x+r} (a\mu)_{x+r}^i \bar{a}_{x+r}^i \, dr$$

$$\doteq {}^zM_{x+1}^{ia} + (1-k)\, {}^zC_x^{ia} = {}^zM_x^{ia} - k\, {}^zC_x^{ia},$$

and let

$$^z\overline{M}^{ia}_x = \int_0^1 {}^zM^{ia}_{x+k}\,dk$$

$$\doteq \tfrac{1}{2}({}^zM^{ia}_x + {}^zM^{ia}_{x+1}) = {}^zM^{ia}_x - \tfrac{1}{2}{}^zC^{ia}_x = {}^zM^{ia}_{x+1} + \tfrac{1}{2}{}^zC^{ia}_x$$

and $$^z\overline{R}^{ia}_x = \sum_{t=0}^{64-x} {}^z\overline{M}^{ia}_{x+t}$$

$$\doteq \tfrac{1}{2}({}^zR^{ia}_x + {}^zR^{ia}_{x+1}) = {}^zR^{ia}_x - \tfrac{1}{2}{}^zM^{ia}_x = {}^zR^{ia}_{x+1} + \tfrac{1}{2}{}^zM^{ia}_x.$$

Then the value of the ill-health pensions to which the members will be entitled in respect of their service between ages $x+t$ and $x+t+1$ is obtained by multiplying $1/m$th of ΣS by

$$\int_0^1 \frac{{}^zM^{ia}_{x+t+k}}{{}^sD_x}\,dk = \frac{{}^z\overline{M}^{ia}_{x+t}}{{}^sD_x}.$$

The factor for valuing the future-service ill-health pensions, when fractions of a year count, is therefore

$$\sum_{t=0}^{64-x} \frac{{}^z\overline{M}^{ia}_{x+t}}{{}^sD_x}$$

$$= \frac{{}^z\overline{R}^{ia}_x}{{}^sD_x}, \tag{29·21}$$

which may be taken as being equal to

$$\frac{\tfrac{1}{2}({}^zR^{ia}_x + {}^zR^{ia}_{x+1})}{{}^sD_x} = \frac{{}^zR^{ia}_x - \tfrac{1}{2}{}^zM^{ia}_x}{{}^sD_x} = \frac{{}^zR^{ia}_{x+1} + \tfrac{1}{2}{}^zM^{ia}_x}{{}^sD_x}. \tag{29·22}$$

For valuing age pensions based on final salary we define ${}^zC^{ra}_x$ in the same way as ${}^zC^{ia}_x$ for all ages up to 64 inclusive but for age 65 we write

$${}^zC^{ra}_{65} = s_{64}\,C^{ra}_{65}.$$

We also write $${}^zM^{ra}_x = \sum_{t=0}^{65-x} {}^zC^{ra}_{x+t}$$

and (if the dates of entry are random in relation to the birthdays)

$${}^zR^{ra}_x = \sum_{t=0}^{64-x} {}^zM^{ra}_{x+t} + \tfrac{1}{2}{}^zM^{ra}_{65}.$$

The functions ${}^zM^{ra}_{x+k}$, ${}^z\overline{M}^{ra}_x$ and ${}^z\overline{R}^{ra}_x$ are defined in the same way as ${}^zM^{ia}_{x+k}$, ${}^z\overline{M}^{ia}_x$ and ${}^z\overline{R}^{ia}_x$, respectively (${}^z\overline{M}^{ra}_{65}$ and ${}^z\overline{R}^{ra}_{65}$ being zero).

The factor for valuing the past-service age pensions is

$$\frac{{}^zM^{ra}_x}{{}^sD_x}. \tag{29·23}$$

The factor for valuing the future-service age pensions, when fractions of a year do not count, is

$$\frac{{}^{z}R^{ra}_{x+1}}{{}^{s}D_x} \tag{29·24}$$

and the factor when fractions of a year count is

$$\frac{{}^{z}\bar{R}^{ra}_{x}}{{}^{s}D_x}. \tag{29·25}$$

In practice the numerator of formula (29·25) will be evaluated in a similar way to the numerator of formula (29·21).

When fractions of a year do not count and only the curtate past service is allowed for in valuing the past-service pensions, it can easily be shown that suitable factors for valuing the future-service pensions (provided that the anniversaries of the entry dates are distributed at random over the calendar year) are

$$\frac{{}^{z}R^{ia}_{x+\frac{1}{2}}}{{}^{s}D_x} \quad \text{and} \quad \frac{{}^{z}R^{ra}_{x+\frac{1}{2}}}{{}^{s}D_x}.$$

It will be noted that these factors are, for practical purposes, identical with formulae (29·21) and (29·25).

If the past service of each member is obtained by adding $\frac{1}{2}$ to his curtate past service, the formulae for valuing the future-service pensions (again provided that the anniversaries of the entry dates are distributed at random over the calendar year) are (29·20) and (29·24) when fractions do not count, or (29·21) and (29·25) when fractions count.

29·8. Pensions based on 'final average' salary

The pension may be based on the average salary during, say, the last 5 years of service. A pension of this kind is sometimes described as being based on 'final average' salary.

The procedure is the same as when the pension is based on the final salary, except that we write

$${}^{z}C^{ia}_{x} \doteq z_{x+\frac{1}{2}}\, C^{ia}_{x}$$

and $${}^{z}C^{ra}_{x} \doteq z_{x+\frac{1}{2}}\, C^{ra}_{x} \quad (\text{but } {}^{z}C^{ra}_{65} = z_{65}\, C^{ra}_{65}),$$

where z_x is defined as being the average salary scale ratio during the five years preceding age x, i.e.

$$z_x = \tfrac{1}{5}(s_{x-5} + s_{x-4} + s_{x-3} + s_{x-2} + s_{x-1})$$

and
$$z_{x+\frac{1}{2}} = \tfrac{1}{5}(\tfrac{1}{2}s_{x-5} + s_{x-4} + s_{x-3} + s_{x-2} + s_{x-1} + \tfrac{1}{2}s_x).$$

This method may need some modification if there has been a general raising of the salary scale, or if individual highly paid members have received substantial increases, during the last five years.

29·9. Maximum and minimum pensions

Sometimes the rules of a fund provide that the pension shall not be less than a certain minimum, say £100, or that salaries in excess of a certain figure shall be treated as being equal to that figure for the purpose of calculating the pension. Provisions of this kind may be dealt with on the lines suggested in section 8 of Chapter 28 but the method is not entirely satisfactory for the reasons stated there.

A very usual rule is that the pension shall be $\frac{1}{60}$th of the salary for each year of service with a maximum of $\frac{40}{60}$ths. The methods of dealing with this rule may be illustrated on the assumption that fractions of a year count.

If the pension is based on average salary, there are two forms of the rule:

(*a*) Salary received more than 40 years before retirement is to be disregarded in calculating the pension. To give effect to this, for a member who entered, say, at age 22 nearest birthday and who is now aged between 25 and 62, we may deduct from the normal factor (to be multiplied by $\frac{1}{60}$th of the salary now coming into force)

$$\frac{s_{22}\overline{M}_{62}^{(i+r)a} + s_{23}\overline{M}_{63}^{(i+r)a} + s_{24}\overline{M}_{64}^{(i+r)a}}{{}^sD_x}.$$

The method may be suitably adjusted for members who are still under age 25 and for those who have served for more than 40 years. It is open to the objection that it involves using the salary scale to estimate past salaries—a purpose for which it may not be suitable.

(*b*) The maximum pension is to be $\frac{40}{60}$ths of the average salary throughout service. This rule may be dealt with by calculating a special commutation function, ${}_yC_{y+40}$, ${}_yC_{y+41}$, ${}_yC_{y+42}$, ... for $y=24, 23, 22, \ldots$. For example, for a member who entered at age 22 nearest birthday, we may write

$$ {}_{22}C_{62}=\frac{40}{40\frac{1}{2}}\,C_{62}^{(i+r)a}, $$

$$ {}_{22}C_{63}=\frac{40}{41\frac{1}{2}}\,C_{63}^{(i+r)a}, $$

$$ {}_{22}C_{64}=\frac{40}{42\frac{1}{2}}\,C_{64}^{(i+r)a} $$

and

$$ {}_{22}C_{65}=\frac{40}{43}\,C_{65}^{ra}. $$

In summing $C^{(i+r)a}$ to obtain $M^{(i+r)a}$ this special C, instead of the ordinary one, will have to be used.

If the pension is based on final salary, care must be taken to limit n to 40 in calculating ΣnS for the past-service pension. In those cases where the past service is 40 years or more there is no future-service pension. Where the past service is less than 40 years, we must deduct

$$ \frac{{}^z\overline{R}_{y+40}^{(i+r)a}}{{}^sD_x} $$

from the future service factor, y being the nearest age at entry.

29·10. Death benefit—past contributions

The death benefit provided by a pension fund is usually a cash payment equal to the employee's contributions with or without interest; it is payable only if he dies before retirement. Let us first assume that compound interest is allowed at rate j (the valuation rate being i) and that the past contributions of each member with interest at rate j up to the valuation date are included in the valuation data.

Then the value of a death benefit equal to the past contributions with interest, in respect of a group of members aged x nearest birthday, is obtained by multiplying their past contributions with interest by

$$ \int_0^{65-x} v^t(1+j)^t\frac{(al)_{x+t}}{(al)_x}(a\mu)_{x+t}^d\,dt. $$

If we write

$$C_x^d = \int_0^1 v^{x+t}(al)_{x+t}(a\mu)_{x+t}^d dt \doteq v^{x+\frac{1}{2}}(ad)_x^d,$$

$$M_x^d = \int_0^{65-x} v^{x+t}(al)_{x+t}(a\mu)_{x+t}^d dt = \sum_{t=0}^{64-x} C_{x+t}^d,$$

$$D_x^{(J)} = D_x \text{ at rate of interest } J,$$

$$C_x^{d(J)} = C_x^d \text{ at rate of interest } J$$

and $$M_x^{d(J)} = M_x^d \text{ at rate of interest } J,$$

and if we define J by the relationship

$$1 + J = \frac{1+i}{1+j},$$

the factor for valuing the death benefit becomes

$$\frac{M_x^{d(J)}}{D_x^{(J)}}. \qquad (29\cdot26)$$

The foregoing development implies the assumption that j is less than i, so that J is positive, but formula (29·26) will not be invalidated if J is negative.

If we write $${}^jM_x^d = (1+j)^{-x} M_x^{d(J)},$$

formula (29·26) can be written

$$\frac{{}^jM_x^d}{D_x}. \qquad (29\cdot27)$$

In practice, formula (29·27) is usually to be preferred to formula (29·26) because, as will be seen in the next section of this chapter, ${}^jM_x^d$ is needed for valuing the death benefit in respect of future contributions, but $D_x^{(J)}$ is not needed for any other purpose.

$C_x^{d(J)}$ will be obtained by the approximation

$$C_x^{d(J)} \doteq v^{x+\frac{1}{2}}(1+j)^{x+\frac{1}{2}}(ad)_x^d,$$

$M_x^{d(J)}$ by summation and ${}^jM_x^d$ by multiplying $M_x^{d(J)}$ by $(1+j)^{-x}$.

When the rate of interest that is allowed on the contributions is identical with the valuation rate i, the calculations are simplified because J becomes zero, $C_x^{d(J)}$ becomes $(ad)_x^d$ and $M_x^{d(J)}$ becomes $(al)_x^d$ (defined as in formula (19·5)). Formula (29·26) therefore becomes

$$\frac{(al)_x^d}{(al)_x} \qquad (29\cdot28)$$

and formula (29·27) becomes

$$\frac{{}^{i}M_x^d}{D_x}, \tag{29·29}$$

where ${}^{i}M_x^d = v^x(al)_x^d$.

When no interest is allowed, j is zero and formulae (29·26) and (29·27) become

$$\frac{M_x^d}{D_x}. \tag{29·30}$$

The employer may not be able to quote the past contributions with interest but he will usually be able to give the past contributions without interest. If the rules of the fund provide for interest on the contributions, a good approximation to the contributions with interest may be obtained by multiplying the contributions without interest by the ratio which the contributions with interest bear to the contributions without interest on the basis of the salary scale. For a member who entered at age y nearest birthday, this ratio is approximately

$$\sum_{t=y}^{x-1} s_t(1+j)^{x-t-\frac{1}{2}} \Big/ \sum_{t=y}^{x-1} s_t = (1+j)^{x-\frac{1}{2}} \sum_{t=y}^{x-1} s_t(1+j)^{-t} \Big/ \sum_{t=y}^{x-1} s_t.$$

If the contributions without interest are not available, a less satisfactory approximation to the contributions with interest may be obtained by multiplying the contributions per annum coming into force on the day after the valuation date by

$$(1+j)^{x-\frac{1}{2}} \sum_{t=y}^{x-1} s_t(1+j)^{-t}/s_x.$$

If necessary, a similar method can be used for obtaining the past contributions without interest.

29·11. Death benefit—future contributions

In order to develop an expression for the value of a death benefit equal to the future contributions with interest at rate j, we may introduce three more commutation functions:

$${}^{j}\overline{M}_x^d = \int_0^1 {}^{j}M_{x+k}^d\, dk \doteq \tfrac{1}{2}({}^{j}M_x^d + {}^{j}M_{x+1}^d),$$

$${}^{sj}\overline{M}_x^d = s_x\, {}^{j}\overline{M}_x^d$$

and

$${}^{sj}\overline{R}_x^d = \sum_{t=0}^{64-x} {}^{sj}\overline{M}_{x+t}^d.$$

Let the total contributions per annum, coming into force on the day after the valuation date, of a group of members aged x nearest birthday be ΣrS. Then the value of the death benefit in respect of the contributions which these members will pay during the year of age $x+t$ to $x+t+1$ is obtained by multiplying ΣrS by

$$\int_0^1 v^{t+k}\frac{s_{x+t}}{s_x}\cdot\frac{(al)_{x+t+k}}{(al)_x}\cdot\frac{{}^jM^d_{x+t+k}}{D_{x+t+k}}dk=\int_0^1\frac{s_{x+t}\,{}^jM^d_{x+t+k}}{{}^sD_x}dk$$
$$=\frac{s_{x+t}\,{}^j\overline{M}^d_{x+t}}{{}^sD_x}$$
$$=\frac{{}^{sj}\overline{M}^d_{x+t}}{{}^sD_x}.$$

The factor for valuing the death benefit in respect of all the future contributions is therefore

$$\sum_{t=0}^{64-x}\frac{{}^{sj}\overline{M}^d_{x+t}}{{}^sD_x}$$
$$=\frac{{}^{sj}\overline{R}^d_x}{{}^sD_x}. \qquad (29\cdot31)$$

When the rate of interest that is allowed on the contributions is equal to i, the formula becomes

$$\frac{{}^{si}\overline{R}^d_x}{{}^sD_x}, \qquad (29\cdot32)$$

where $${}^i\overline{M}^d_x=\int_0^1 {}^iM^d_{x+t}dt \doteq \tfrac{1}{2}({}^iM^d_x+{}^iM^d_{x+1}),$$
$${}^{si}\overline{M}^d_x=s_x\,{}^i\overline{M}^d_x$$

and $${}^{si}\overline{R}^d_x=\sum_{t=0}^{64-x}{}^{si}\overline{M}^d_{x+t}.$$

When no interest is allowed, the formula becomes

$$\frac{{}^s\overline{R}^d_x}{{}^sD_x}, \qquad (29\cdot33)$$

where $$\overline{M}^d_x=\int_0^1 M^d_{x+t}dt \doteq \tfrac{1}{2}(M^d_x+M^d_{x+1}),$$
$${}^s\overline{M}^d_x=s_x\overline{M}^d_x$$

and $${}^s\overline{R}^d_x=\sum_{t=0}^{64-x}{}^s\overline{M}^d_{x+t}.$$

29·12. Withdrawal benefit

Many pension funds provide for a benefit equal to the employee's contributions with or without interest (usually without interest) in the event of the employee's withdrawal from the service of the employer otherwise than by retirement on pension. The procedure for valuing this return is exactly the same as that for valuing the death benefit; the commutation functions C_x^d, M_x^d, etc., in the development of the formula for returns on death will be replaced by C_x^w, M_x^w, etc.

29·13. Miscellaneous benefits

Some funds provide for a lump sum, in addition to a pension, on retirement. Others provide for a lump sum on death, or for the contributions with interest, payable on death, to be increased to a fixed amount if they fall short of it. In many funds the pension is payable for a minimum period of, say, 5 years or there is a decreasing lump sum benefit on death if this occurs within a short period after retirement. Benefits of this kind can be dealt with by modifications of the formulae already discussed.

29·14. Calculation of rates of contribution

The calculation of the appropriate rates of contribution to be paid in respect of new entrants is a simple matter. For example, let us suppose that employer and employee contribute equally (the employer paying at the end of each year), that the pension benefit is $\frac{1}{60}$th of the final salary for each completed year of service with a maximum of $\frac{40}{60}$ths, that the death benefit is equal to the employee's contributions with interest and that the withdrawal benefit is equal to the employee's contributions without interest. Then if the employee's contribution is to be r per cent of his salary, we have the following equation to determine r for a member entering at exact age 20.

$$\cdot 01r(1+v^{\frac{1}{2}})\,{}^s\bar{N}_{20} = \tfrac{1}{60}\{{}^zR_{21}^{(i+r)a} - {}^zR_{61}^{(i+r)a}\} + \cdot 01r({}^{sj}\bar{R}_{20}^d + {}^s\bar{R}_{20}^w).$$

Strictly speaking, if salaries are revised yearly on a fixed date such as 1 January, the foregoing equation is not quite correct because, unless the new entrant joins on a salary revision date, the incidence of his salary increases will not agree with that allowed for in the formulae used in the valuation. However, as the rate is always 'rounded off', refinements in the calculation are unnecessary.

Expenses are not included in the equation as they are usually borne by the employer.

In some funds the rate of contribution paid by the employee and employer is not sufficient, for all ages at entry, to provide the benefits. The admission of new entrants at ages where the contribution is inadequate will place a strain on the fund which may have to be offset by special contributions by the employer.

EXAMPLES

Example 29·1

On the basis of the rates of death, retirement and withdrawal and the salary scale reproduced on pages 246 and 248 of this book and assuming 3 per cent interest, value the following benefits to a group of employees aged 40 with total salaries of £10,000 coming into force on the day after the valuation date.

(*a*) Ill-health and retirement pensions of 2 per cent of the five-year final average salary for each year of service, only completed years counting. (The sum of the products obtained by multiplying each member's nearest integral past service by his new salary is £100,000.)

(*b*) Return of employees' contributions without interest on death. (Employees' contributions are 3 per cent of salary, and past contributions amount to £2,400.)

Solution

(*a*) The factor for valuing the past-service pensions is

$$\frac{{}^{z}M_x^{ia}}{{}^{s}D_x}+\frac{{}^{z}M_x^{ra}}{{}^{s}D_x}$$

and the factor for valuing the future-service pensions when only completed years count is

$$\frac{{}^{z}R_{x+1}^{ia}}{{}^{s}D_x}+\frac{{}^{z}R_{x+1}^{ra}}{{}^{s}D_x}.$$

${}^z R^{ia}_{41}$ is not in the tables but ${}^z\bar{R}^{ia}_{40}$ is tabulated. Now

$${}^z\bar{R}^{ia}_{40} \doteq \tfrac{1}{2}{}^z R^{ia}_{40} + \tfrac{1}{2}{}^z R^{ia}_{41} = {}^z R^{ia}_{41} + \tfrac{1}{2}{}^z M^{ia}_{40}.$$

$$\therefore \quad {}^z R^{ia}_{41} \doteq {}^z\bar{R}^{ia}_{40} - \tfrac{1}{2}{}^z M^{ia}_{40}$$

and similarly $\quad {}^z R^{ra}_{41} \doteq {}^z\bar{R}^{ra}_{40} - \tfrac{1}{2}{}^z M^{ra}_{40}.$

Hence, the factor for valuing the past-service pensions is

$$\frac{77{,}064 + 346{,}377}{2{\cdot}700 \times 24{,}758} = 6{\cdot}33,$$

giving a value of £100,000 × ·02 × 6·33 = £12,660, and the factor for valuing the future-service pensions is

$$\frac{1{,}210{,}937 + 7{,}980{,}000}{2{\cdot}700 \times 24{,}758} - \tfrac{1}{2} \times 6{\cdot}33 = 137{\cdot}49 - 3{\cdot}16 = 134{\cdot}33,$$

giving a value of £10,000 × ·02 × 134·33 = £26,866. The total value of the pensions is therefore £39,526, approximately.

(*b*) The factor for valuing the death benefit in respect of past contributions is

$$\frac{M^d_x}{D_x} = \frac{2{,}780}{24{,}758} = {\cdot}112,$$

giving a value of £2,400 × ·112 = £269.

The factor for valuing the death benefit in respect of future contributions is

$$\frac{{}^s\bar{R}^d_x}{{}^sD_x} = \frac{120{,}856}{2{\cdot}700 \times 24{,}758} = 1{\cdot}808,$$

giving a value of £10,000 × ·03 × 1·808 = £542.

The total value of the return of contributions on death is therefore £811.

Example 29·2

On the assumptions set out in the previous example, find the value for the same group of employees of an age retirement pension which provides the undernoted percentages of average annual salary throughout service for each complete year of service.

Retirement age		Percentage
Not less than	Less than	
60	63	1·5
63	64	1·6
64	65	1·7
65 exact		1·8

Total complete past-service salaries £80,000.

Solution

If retirement occurs in the year of age x to $x+1$, the factor (to be multiplied by the past-service salaries) for obtaining the value of the past-service pensions is

$$C_x^{ra'} = C_x^{ra} \times r_x,$$

where r_x is the percentage appropriate to age x to $x+1$.

The factor for valuing the whole past-service age pensions is therefore

$$\frac{M_x^{ra'}}{D_x},$$

where $$M_x^{ra'} = \sum_{t=0}^{65-x} C_{x+t}^{ra'}.$$

The factor for the future-service age pensions is, by a similar development to that used to obtain formula (29·18),

$$\frac{{}^sR_{x+1}^{ra'}}{{}^sD_x}.$$

From the tables we have

x	C_x^{ra}	Percentage	$100C_x^{ra'}$	$100M_x^{ra'}$	s_{x-1}	$100{}^sM_x^{ra'}$	$100{}^sR_x^{ra'}$
60	25,429	1·5	38,144	157,924	3·574	564,420	2,097,696
61	13,856	1·5	20,784	119,780	3·600	431,208	1,533,276
62	7,097	1·5	10,645	98,996	3·624	358,762	1,102,068
63	5,770	1·6	9,232	88,351	3·646	322,128	743,306
64	4,665	1·7	7,931	79,119	3·666	290,050	421,178
65	39,549	1·8	71,188	71,188	3·684	131,128	131,128

$$100M_{41}^{ra'} = 100M_{42}^{ra'} = \ldots = 100M_{60}^{ra'} = 157{,}924.$$

Now

$$\sum_{x=40}^{58} s_x = -\tfrac{1}{2}s_{39} + 5(z_{44\frac{1}{2}} + z_{49\frac{1}{2}} + z_{54\frac{1}{2}} + z_{59\frac{1}{2}}) - \tfrac{1}{2}s_{59}$$

$$= 5(2{\cdot}793 + 3{\cdot}078 + 3{\cdot}313 + 3{\cdot}498) - 1{\cdot}317 - 1{\cdot}787$$

$$= 63{,}410 - 3{\cdot}104$$

$$= 60{\cdot}306.$$

Hence $$100\,{}^sR_{41}^{ra'} = 60{\cdot}306 \times 157{,}924 + 2{,}097{,}696$$

$$= 11{,}621{,}461.$$

The value of the past-service pensions is

$$80{,}000 \times \frac{1{,}579{\cdot}24}{24{,}758} = £5{,}103$$

and the value of the future-service pensions is

$$10{,}000 \times \frac{116{,}214{\cdot}61}{2{\cdot}700 \times 24{,}758} = £17{,}385,$$

giving a total value of £22,488, approx.

CHAPTER 30

WIDOWS' AND ORPHANS' FUNDS

30·1. Introduction

An ordinary pension fund provides for the employee on his retirement owing to age or ill health; a widows' and orphans' fund provides for the employee's dependants on his death. Widows' pensions usually cease on remarriage. While widows' and orphans' benefits may form part of a pension fund they are more usually provided by a separate fund. In this chapter we shall consider first the valuation of widows' benefits and then the valuation of orphans' benefits.

There are two alternative methods of valuing widows' pensions. These are known as the collective method and the reversionary method, respectively. Briefly, the principle of the collective method is to group together all members aged x nearest birthday regardless of marital status and to introduce a proportion married at each age when calculating the factors. Under the reversionary method the members aged x are sub-grouped according to their marital status and the married may be further sub-grouped according to the age of the wife; the benefits payable to the present wives of the married men are valued by reversionary annuities and special factors are constructed for valuing benefits to second wives of the married men and future wives of the unmarried. It is not proposed to comment here on the relative merits and demerits of the two methods, nor is it possible, within the scope of this chapter, to deal exhaustively with the various types of benefit that may be encountered in practice and the different ways of valuing them. All that will be attempted will be to give a general idea of the two methods with the aid of some fairly simple and representative formulae. These may be regarded as forming a basis for building up suitable formulae to meet the conditions of any particular fund.

30·2. Widows' pensions—collective method

By the collective method, the factor for valuing a pension of 1 per annum payable to each widow in respect of a group of men now aged x nearest birthday may be written

$$\int_0^\infty v^t \frac{l_{x+t}}{l_x} \mu_{x+t} h_{x+t} \bar{a}''_{y+t} dt$$

$$\doteqdot \sum_{t=0}^{\infty} \frac{v^{x+t+\frac{1}{2}} d_{x+t}}{D_x} h_{x+t+\frac{1}{2}} \bar{a}''_{y+t+\frac{1}{2}}. \tag{30·1}$$

In this expression d_{x+t} and D_x are based on a mortality table appropriate to male members irrespective of marital status, $h_{x+t+\frac{1}{2}}$ is the proportion married at age $x+t+\frac{1}{2}$ and $\bar{a}''_{y+t+\frac{1}{2}}$ is the weighted average value of 1 per annum, ceasing on death or remarriage, to the widow of a man who has just died at age $x+t+\frac{1}{2}$. Strictly speaking the proportion married should be the proportion married *among those who die* between ages $x+t$ and $x+t+1$, not among those who are living at age $x+t+\frac{1}{2}$. For simplicity the suffix $y+t+\frac{1}{2}$ is used in this and similar formulae to represent the average age of the wives of husbands aged $x+t+\frac{1}{2}$; this is not necessarily equal to y increased by $t+\frac{1}{2}$, where y is the average age of the wives of husbands aged x.

Formula (30·1) is based on the assumption that the benefit is not related in any way to the member's service. The rules governing widows' pensions often stipulate that marriages after the member's retirement shall not count. In these circumstances formula (30·1) becomes

$$\int_0^{65-x} v^t \frac{(al)_{x+t}}{(al)_x} (a\mu)^d_{x+t} h_{x+t} \bar{a}''_{y+t} dt$$

$$+ \int_0^{65-x} v^t \frac{(al)_{x+t}}{(al)_x} (a\mu)^r_{x+t} h_{x+t} \bar{a}''_{x+t|y+t} dt$$

$$+ v^{65-x} \frac{(al)_{65}}{(al)_x} h_{65} \bar{a}''_{65|y+65-x},$$

where r includes ill-health retirements, $\bar{a}''_{x+t|y+t}$ is the weighted average value of a reversionary annuity of 1, ceasing on death or remarriage, to the wife of a man who has just retired at age $x+t$ and $\bar{a}''_{65|y+65-x}$ is a similar annuity to the wife of a man who has just retired at age 65. This formula may be written

$$\frac{M_x^{d(wa)} + {}^rM_x^{d(wa)}}{D_x}, \tag{30·2}$$

where $$C_x^{d(wa)} \doteq v^{x+\frac{1}{2}}(ad)_x^d h_{x+\frac{1}{2}} \bar{a}''_{y+\frac{1}{2}},$$

$$M_x^{d(wa)} = \sum_{t=0}^{64-x} C_{x+t}^{d(wa)},$$

$${}^rC_x^{d(wa)} \doteq v^{x+\frac{1}{2}}(ad)_x^r h_{x+\frac{1}{2}} \bar{a}''_{x+\frac{1}{2}|y+\frac{1}{2}} \quad \text{for} \quad x < 65,$$

$${}^rC_{65}^{d(wa)} = v^{65}(al)_{65} h_{65} \bar{a}''_{65|y+65-x}$$

and $${}^rM_x^{d(wa)} = \sum_{t=0}^{65-x} {}^rC_x^{d(wa)}.$$

The similarity between M_x^{ia} and $M_x^{d(wa)}$, and between M_x^{ra} and ${}^rM_x^{d(wa)}$, will be noted.

Formula (30·2) is suitable for valuing the widow's pension if it is fixed in amount. If it is $1/m$th of the husband's average or final salary for each year of his service, formulae of the type ${}^s\bar{R}/{}^sD$, ${}^z\bar{R}/{}^sD$, etc., may be built up from $M_x^{d(wa)}$ and ${}^rM_x^{d(wa)}$ in exactly the same way in which formulae for valuing the member's pension have been built up in Chapter 29 from M_x^{ia} and M_x^{ra}. In some funds the widow's pension is proportionate to the husband's period of service; to obtain suitable formulae for these it is only necessary to omit the salary ratios.

A fund which provides for widows' pensions may also provide for a benefit equal to the member's contributions with interest when a member dies unmarried. This may be valued on the lines indicated in sections 10 and 11 of Chapter 29, C_x^d being summed to obtain M.

The valuation of the contributions is usually a simple matter. Sometimes the married are required to contribute twice as much as the unmarried; the formula will then be

$$\sum_{t=0}^{64-x} \frac{{}^s\bar{D}_{x+t}}{{}^sD_x}(1 + h_{x+t+\frac{1}{2}})$$

if the contributions depend on salary, or

$$\sum_{t=0}^{64-x} \frac{\bar{D}_{x+t}}{D_x}(1 + h_{x+t+\frac{1}{2}})$$

if they do not. The proportion married that is used here must relate to the living, not to those who die.

30·3. Widows' pensions—reversionary method

By the reversionary method, the value of a pension of 1 per annum to the present wife aged y of a member aged x is

$$
{}^{m}B_{xy} = \int_0^\infty v^t \frac{l_{x+t}}{l_x} \cdot \frac{l_{y+t}}{l_y} \mu_{x+t} \bar{a}'_{y+t} dt
$$

$$
\doteq \sum_{t=0}^{\infty} \frac{d_{x+t}}{l_x} \cdot \frac{D_{y+t+\frac{1}{2}}}{D_y} \bar{a}'_{y+t+\frac{1}{2}}, \tag{30·3}
$$

where d_{x+t} and l_x come from a table suitable for married male members and $D_{y+t+\frac{1}{2}}$ and D_y come from a table appropriate to wives; $\bar{a}'_{y+t+\frac{1}{2}}$ is an annuity ceasing on death or remarriage. The formula may be applied either to the whole group of married members who are aged between $x-\frac{1}{2}$ and $x+\frac{1}{2}$, in which event y will represent the average age of their wives, or to the sub-group of married members who are aged between $x-\frac{1}{2}$ and $x+\frac{1}{2}$ and whose wives are aged between, say $y-1\frac{1}{2}$ and $y+1\frac{1}{2}$.

The factor for valuing the liability in respect of future first wives of bachelors aged x is

$$
{}^{m}_{b}B_x \doteq \sum_{t=0}^{\infty} \frac{v^{x+t+\frac{1}{2}} (bd)^h_{x+t}}{v^x (bl)_x} \cdot {}^{m}B_{[x+t+\frac{1}{2}]:y+t+\frac{1}{2}}, \tag{30·4}
$$

where $(bd)^h_{x+t}$ and $(bl)_x$ come from a combined mortality and marriage table and the ratio $(bd)^h_{x+t}/(bl)_x$ is therefore the probability that a bachelor aged x will marry between ages $x+t$ and $x+t+1$; ${}^{m}B_{[x+t+\frac{1}{2}]:y+t+\frac{1}{2}}$ is the weighted average value of a reversionary pension of 1 per annum to the wife of a newly married man aged $x+t+\frac{1}{2}$.

For valuing the liability in respect of future wives of widowers, the ratio $(bd)^h_{x+t}/(bl)_x$ in formula (30·4) should, strictly speaking, be replaced by the corresponding probability in respect of a widower. In practice, it may be considered sufficient to use formula (30·4) as it stands.

The factor in respect of future wives of married men may be written

$$
\sum_{t=0}^{\infty} \frac{v^{x+t+\frac{1}{2}} (hd)^w_{x+t}}{v^x (hl)_x} \cdot {}^{m}_{b}B_{x+t+\frac{1}{2}}, \tag{30·5}
$$

where $(hd)^w_{x+t}$ and $(hl)_x$ come from a double-decrement mortality and widowerhood table and $(hd)^w_{x+t}$ is the number of married men becoming widowers between ages $x+t$ and $x+t+1$; ${}^m_bB_{x+t+\frac{1}{2}}$ should, strictly speaking, be a factor appropriate to a man who has just become a widower.

The liability in respect of subsequent marriages could be represented by similar formulae, but this liability is usually so small that a rough estimate of it is sufficient.

If the contributions do not vary with marital status and if they cease, for example, on the attainment of age 65 by the husband, they may be valued by

$$\bar{a}_{x:\overline{65-x}|} \tag{30·6}$$

from a table applicable to all males.

If the amount of the contribution depends upon marital status, the contribution of a married man may be regarded as consisting of two portions—the basic contribution which would be payable if he were unmarried and the additional contribution for a married man. The basic contributions of all the men aged x can be valued by formula (30·6). The formula for valuing the additional contributions of the married men during their present marriages is

$$\sum_{t=0}^{64-x} \frac{l_{x+t+\frac{1}{2}}}{l_x} \cdot \frac{D_{y+t+\frac{1}{2}}}{D_y}, \tag{30·7}$$

where the functions come from the tables used for formula (30·3). The formula for valuing the additional contributions during the future married status of bachelors takes the form

$$\sum_{t=0}^{64-x} \frac{v^{x+t+\frac{1}{2}}(bd)^h_{x+t}}{v^x(bl)_x} \bar{a}_{[x+t+\frac{1}{2}]:y+t+\frac{1}{2}}, \tag{30·8}$$

where $\bar{a}_{[x+t+\frac{1}{2}]:y+t+\frac{1}{2}}$ is the weighted average value of an annuity of 1 payable during the joint lifetime of a newly married man aged $x+t+\frac{1}{2}$ and his wife. This formula, like formula (30·4), is, strictly speaking, not suitable for widowers, but it may be considered sufficiently accurate for them in practice.

The value of the benefit on death unmarried is usually small and it is normally calculated by the collective method.

The reversionary method can be extended to the valuation of pensions which are related to the husband's service but the formulae are complicated and it is not proposed to consider them here. The reader who is interested is referred to *Pension and Widows' and Orphans' Funds* by R. J. W. Crabbe and C. A. Poyser.

30·4. Orphans' benefits

Orphans' benefits are often included in a scheme which provides for pensions to widows. The liability for orphans' benefits is usually comparatively small and refinements in their valuation are therefore out of place.

The orphans' benefit is usually an annuity ceasing at a fixed age. It is customary to ignore the mortality of orphans and to take the value of an annuity ceasing at age k in respect of a child aged c as being

$$\bar{a}_{\overline{k-c}|}.$$

Numerous methods of valuing these benefits are available. One is to use the formula

$$\sum_{t=0}^{\infty} \frac{v^{x+t+\frac{1}{2}} d_{x+t}}{D_x} \{h_{x+t+\frac{1}{2}}(Oa)_{x+t+\frac{1}{2}} + h'_{x+t+\frac{1}{2}}(Oa)'_{x+t+\frac{1}{2}}\}, \qquad (30·9)$$

where h' is the proportion of widowers, (Oa) is the 'average family annuity', or average value of the orphans' benefit, in respect of a man who has just died married and $(Oa)'$ is the corresponding function for a widower who has just died. This formula would be applied to all males aged x, regardless of marital status.

Another method is to use

$$\sum_{t=0}^{\infty} \frac{D_{x+t+\frac{1}{2}}}{D_x} i_{x+t} \bar{a}'_{x+t+\frac{1}{2}|0}, \qquad (30·10)$$

where i_{x+t} is the central issue rate (or approximately the force of issue at age $x+t+\frac{1}{2}$) and $\bar{a}'_{x+t+\frac{1}{2}|0}$ is a reversionary annuity, ceasing at age k, to a child aged 0. This formula also would be applied to all males. It allows only for children not yet born; benefits to existing children would have to be valued by reversionary annuities of the type $\bar{a}'_{x|c}$.

EXAMPLES

Example 30·1

(*a*) On the basis of the rates of death, retirement and withdrawal reproduced on pages 246 and 248 of this book and assuming 3 per cent interest, calculate the value of pensions of £100 per annum to the widows of N lives aged 60, if the pensions are payable only in the event of death while in active service. The following table gives the proportion married and the value of a widow's annuity at each age:

Age of husband	Proportion married	Mean value of widow's annuity if death occurs during year of age
60–61	·75	13·0
61–62	·75	12·5
62–63	·74	12·0
63–64	·74	11·5
64–65	·73	11·0

(*b*) On the same basis and assuming the salary scale on page 248, what would be the value of widows' pensions of 25 per cent of final five-year average salary if the lives have total salaries of S?

Solution

(*a*) The value of the widows' pensions is

$$N\times\frac{100}{D_{60}}\sum_{x=60}^{64} C_x^d h_{x+\frac{1}{2}}\bar{a}''_{y+\frac{1}{2}}.$$

Age x	C_x^d	$h_{x+\frac{1}{2}}$	$\bar{a}''_{y+\frac{1}{2}}$	$C_x^d h_{x+\frac{1}{2}}\bar{a}''_{y+\frac{1}{2}}$
60	151	·75	13·0	1,472
61	129	·75	12·5	1,209
62	119	·74	12·0	1,057
63	113	·74	11·5	962
64	109	·73	11·0	875
				5,575

Hence, the value of the pensions is

$$N\times\frac{100\times 5{,}575}{10{,}725}=N\times £51{\cdot}98.$$

(*b*) The value of the widows' pensions is

$$S\times\frac{\cdot 25}{{}^sD_{60}}\sum_{x=60}^{64} C_x^d z_{x+\frac{1}{2}}h_{x+\frac{1}{2}}\bar{a}''_{y+\frac{1}{2}}.$$

Age x	$C_x^d h_{x+\frac{1}{2}} \bar{a}''_{y+\frac{1}{2}}$	$z_{x+\frac{1}{2}}$	$C_x^d z_{x+\frac{1}{2}} h_{x+\frac{1}{2}} \bar{a}''_{y+\frac{1}{2}}$
60	1,472	3·529	5,195
61	1,209	3·558	4,302
62	1,057	3·585	3,789
63	962	3·610	3,473
64	875	3·633	3,179
			19,938

Hence, the value of the pensions is

$$S\times\frac{\cdot 25\times 19{,}938}{3\cdot 600\times 10{,}725}=S\times £\cdot 129.$$

Example 30·2

Show how you would calculate the value of pensions payable only on death while in service and before age 65 to the widows of married men all aged 50, where the average age of the wives is also 50. The calculation is to be limited to pensions to present wives and it may be assumed

(*a*) that the mortality of the male lives follows Gompertz's law,

(*b*) that the mortality of the female lives is the same as that of male lives 4 years younger,

(*c*) that the force of withdrawal of male lives is equal to the force of mortality,

(*d*) that there will be no divorces,

(*e*) that for the age range 50 to 65, the value of a widow's pension (ceasing on remarriage) is 10·0.

Solution

At age $50+t$ the various decrements from the double-decrement joint-life table for husbands and wives are:

Decrement	Corresponding force of mortality in the Gompertz Table for male lives
Death of the male life	μ_{50+t}
Death of the wife	μ_{46+t}
Withdrawal of the male life	μ_{50+t}

If $2\mu_{50}+\mu_{46}=\mu_x$ the value of a widow's pension of 1 per annum to the present wife of a man aged 50 is

$$\frac{1}{l_x}\int_0^{15} v^t l_{x+t}\mu_{50+t}\, 10{\cdot}0\, dt = \frac{10{\cdot}0}{l_x}\cdot\frac{\mu_{50}}{2\mu_{50}+\mu_{46}}\int_0^{15} v^t l_{x+t}\mu_{x+t}\, dt$$

$$= \frac{10\mu_{50}}{2\mu_{50}+\mu_{46}}\bar{A}^{1}_{x:\overline{15}|}.$$

CHAPTER 31

DISABILITY BENEFITS

31·1. Introduction

A disability benefit is an additional benefit which is sometimes included in a life assurance policy and which, as its name implies, is payable only if the life assured suffers from some form of disability. Incorporation of the benefit in a policy is usually optional and is subject to the payment of an additional premium. The benefit is not normally granted apart from a life policy.

As there are various types and degrees of disability, it is important that any disability clause which is incorporated in a policy should include a carefully worded statement of the conditions in which the benefit will be payable. Disability may be partial (for example, incapacity to follow the life assured's usual occupation) or total. It may also be temporary or permanent. In order to keep the cost to a minimum it is usual to restrict the disability which is covered in connexion with a life policy to 'total and permanent disability'. This is sometimes defined as 'incapacity to follow any occupation whatever for remuneration or profit'. The loss of two limbs, or the loss of sight of both eyes, or the loss of one limb and the loss of sight of one eye, is usually regarded as sufficient evidence of total and permanent disability. Should the life assured suffer from some other form of disability, it will be necessary for him to satisfy the office that it is total and permanent before he can claim the benefit. As recovery may take place even from disability which is admitted by the office to be total and permanent, it is usual to include in the disability clause a provision for production of evidence from time to time that the disability has not ceased and for cessation of the benefit should recovery take place.

The most usual forms of disability benefit are

(i) waiver of premiums during disability,

(ii) waiver of premiums plus an income benefit during disability,

(iii) immediate payment of the sum assured on disablement.

In (i) and (ii), payment of the sum assured is not affected by the disability clause; for example, if the policy is an endowment assurance the sum assured will be payable on the maturity date or previous death, as it would have been if no disability clause had been included. In (iii), however, once the sum assured has been paid on disablement, no further payment is made under the policy.

It may be remarked here that many writers have used the word 'disability' indiscriminately to mean either 'the state of being disabled' or 'the event of becoming disabled'. As this practice may lead to confusion, it is proposed in this book to restrict the use of the word 'disability' to 'the state of being disabled' and to use the word 'disablement' for 'the event of becoming disabled'.

In practice disability benefits are always restricted to disablement which occurs before some age such as 60. This is done in order to provide an attractive benefit at the lowest cost. As some form of disability, or incapacity for work, is a natural characteristic of advancing years, it would be very difficult to find a definition of disability which would severely limit the claims at the higher ages without causing dissatisfaction to the assured.

It will be seen that disability is similar to sickness. The technique which is used for valuing sickness benefits could therefore be adapted to disability. Owing, however, to the fact that disability benefits are not normally granted unless the disability is permanent, a different method, similar to that which is used in connexion with pension funds, has been developed and it will be described in this chapter.

31·2. Notation

There has been much variation in the notation used for the functions which are required for valuing disability benefits. The Actuarial Society of America adopted an 'official' notation for the purpose in 1912 and this is set out in Volume **33** [1932] of the Transactions of the Society. The notation is based on the subdivision of the l_x column of the mortality table into two sections, l_x^{aa} (representing the lives who are not disabled) and l_x^{ii} (representing those who are disabled). The resulting table is known as a combined mortality and disability table (see section 9 of Chapter 19). The letter a stands for 'active' and the letter i for 'invalid'. The American notation was well suited to a method of valuation known as the Hamza method, which will be referred to briefly in section 9 of this chapter, but the modern type of disability benefit requires treatment on pension fund lines. It is therefore proposed in this chapter to use a notation which conforms as closely as possible to that which has already been used in this book for multiple decrement tables and pension funds. In addition the symbol $\bar{a}_x^{ai}$ will be used for the value of a continuous annuity of 1 payable while a life now aged x and active is alive but disabled. This is in agreement with the symbol a_x^{ai} which is used in the American Notation for the corresponding annuity payable yearly.

31·3. The valuation of disability benefits—ordinary whole-life assurances

Let us suppose that we have a double decrement table containing the decrements death and permanent disablement. Let

$(al)_x$ = the number of lives expected to be alive and active at age x out of $(al)_\alpha$ alive at age α, the earliest age in the table,

$(ad)_x^d$ = the expected decrement by death between ages x and $x+1$,

$(ad)_x^i$ = the expected decrement by disablement between ages x and $x+1$,

$(a\mu)_x^i$ = the force of disablement at age x,

$(aq)_x^i = (ad)_x^i/(al)_x$, i.e. the probability that an active life aged x will become disabled before attaining age $x+1$,

$\bar{a}_{[x]}^i$ = the value of a continuous annuity of 1 to a life aged x who has just become disabled.

Then
$$\bar{a}_x^{ai} = \int_0^\infty v^t \frac{(al)_{x+t}}{(al)_x} (a\mu)_{x+t}^i \bar{a}_{[x+t]}^i dt. \tag{31·1}$$

In this formula the upper limit of integration is shown as infinity, it being understood that if, for example, disablement occurring above age 60 does not qualify for benefit, $(a\mu)_x^i$ will be regarded as being zero above that age.

If we write $$(aD)_x = v^x (al)_x,$$

$$(aC)_x^{ia} = \int_0^1 v^{x+t}(al)_{x+t}(a\mu)_{x+t}^i \bar{a}_{[x+t]}^i \, dt$$

$$\doteq (aC)_x^i \, \bar{a}_{[x+\frac{1}{2}]}^i$$

and $$(aM)_x^{ia} = \int_0^\infty v^{x+t}(al)_{x+t}(a\mu)_{x+t}^i \bar{a}_{[x+t]}^i \, dt$$

$$= \sum_{t=0}^{\infty} (aC)_{x+t}^{ia},$$

formula (31·1) becomes $$\bar{a}_x^{ai} = \frac{(aM)_x^{ia}}{(aD)_x}. \tag{31·2}$$

Formula (31·2) is suitable for valuing a waiver of premium benefit in respect of a whole-life assurance with premiums payable throughout life. The use of the continuous annuities $\bar{a}_{[x+t]}^i$ and $\bar{a}_{[x+\frac{1}{2}]}^i$ is suitable even when the premium is payable yearly, because the first premium which is waived may fall due at any time within one year after the date of disablement. It is usual to apply the formula to the office premium, although there may be some slight saving in renewal expenses when premiums are waived.

Formula (31·2) may also be used for valuing an income benefit during disability. Strictly speaking the annuity $\bar{a}_{[x+\frac{1}{2}]}^i$ should be replaced by an annuity which represents correctly the manner and frequency of payment of the benefit. For example, if the benefit is payable monthly in arrear without proportion to the date of death, the annuity should be altered to

$$a_{[x+\frac{1}{2}]}^{i(12)}.$$

If the disability benefit consists of payment of the sum assured (say S) immediately on disablement, this is equivalent to a continuous annuity of $P'' + \delta S$ payable during disability, where P'' is the office premium for the life assurance. Accordingly, formula (31·2) is again suitable.

31·4. The valuation of disability benefits—whole-life assurances subject to limited premiums and endowment assurances

In the case of whole-life assurances subject to limited premiums formula (31·2) is suitable for valuing an income benefit during disability or for valuing δS if the sum assured is payable on disablement. For valuing the waiver of premiums, however, it is necessary to replace

$$\bar{a}^{i}_{[x+\frac{1}{2}]} \quad \text{by} \quad \bar{a}^{i}_{[x+\frac{1}{2}]:\overline{u-x-1}|},$$

where u is the age at which the premiums cease (i.e. the last yearly premium falls due at age $u-1$).

A similar alteration has to be made for valuing all three benefits in the case of an endowment assurance maturing at age u.

31·5. Waiting periods

It is usual to provide that disability must have lasted for a certain period, say 6 months, before a disability benefit will be paid. If the double decrement table is constructed on the basis that the decrement 'disablement' takes effect at the end of the waiting period, the formulae deduced above will require no modification. If, however, the 'disablement' of the double decrement table represents the beginning of the waiting period, the annuities

$$\bar{a}^{i}_{[x+\frac{1}{2}]}, \quad a^{i(12)}_{[x+\frac{1}{2}]} \quad \text{and} \quad \bar{a}^{i}_{[x+\frac{1}{2}]:\overline{u-x-1}|}$$

will need to be adjusted to allow for the absence of payments during the waiting period.

31·6. Net and office premiums for disability benefits

If the premium for the life assurance is payable throughout life, or up to a later age than the limiting age after which disablement does not quality for benefit, it is usual for the disability premium to cease at that age. For example, if the limiting age for disablement is 60, the net yearly premium for a disability benefit will be obtained by replacing $(aD)_x$ in the denominator of the formula by

$$(aN)_x-(aN)_y$$

if the life assurance premiums cease at age y, <60, or by

$$(aN)_x-(aN)_{60}$$

if the life assurance premiums are payable throughout life or cease at or after age 60. If desired, a column of values of $(aN)_x$ could be constructed by summing $(aD)_x$ upwards from age 59; $(aN)_{60}$ would then be zero

The office yearly premium is usually obtained from the net yearly premium by adding a 'percentage' and a 'constant' loading.

When premiums are payable more frequently than yearly it is customary to add the same arbitrary percentage as is added to the life assurance premium. A special point calls for notice, however, when the benefit is a waiver of premiums and the life assurance premiums are not payable for the whole of life. The annuity $\bar{a}^i_{[x+\frac{1}{2}]:\overline{u-x-1}|}$, mentioned in section 4 of this chapter, is suitable when premiums are payable yearly but is, strictly speaking, inadequate when they are payable more frequently. It is usual to make no specific allowance for this point, but to regard it as being covered by the normal loadings.

31·7. Use of a single decrement table

Formulae (31·1) and (31·2) may be written

$$\bar{a}^{ai}_x \doteq \sum_{t=0}^{\infty} \frac{v^{\frac{1}{2}}(aD)_{x+t}(aq)^i_{x+t}\bar{a}^i_{[x+t+\frac{1}{2}]}}{(aD)_x}. \qquad (31\cdot 3)$$

If tables of $(aq)^i_x$ and $\bar{a}^i_{[x]}$ are available, a rough approximation to $\bar{a}^{ai}_x$ may be obtained without the use of a double decrement table by taking the values of D_x from a single decrement table and writing

$$\bar{a}^{ai}_x \doteq \sum_{t=0}^{\infty} \frac{v^{\frac{1}{2}}D_{x+t}(aq)^i_{x+t}\bar{a}^i_{[x+t+\frac{1}{2}]}}{D_x}. \qquad (31\cdot 4)$$

Since D_{x+t}/D_x will be greater than $(aD)_{x+t}/(aD)_x$ for all values of t, this method overstates the value of $\bar{a}^{ai}_x$.

It has, of course, no theoretical justification and it is not suitable for use unless the values of (aq) and $\bar{a}$ are so arbitrary that it is felt that the use of a double-decrement table would be an unnecessary refinement.

31·8. Reserves

In Great Britain, it is customary to hold an arbitrary reserve, equal to say half a year's disability premiums, for future disability benefits to lives who are still active. If desired, a more accurate reserve may be calculated by a formula of the type

$$K\frac{(aM)^{ia}_{x+t}}{(aD)_{x+t}} - P^D_x\frac{(aN)_{x+t}-(aN)_u}{(aD)_{x+t}}, \qquad (31\cdot 5)$$

where K is the disability benefit per annum, P_x^D is the disability premium, $x+t$ is the valuation age and u is the age at which the disability premiums cease.

Benefits to disabled lives should, strictly speaking, be valued by functions of the form $\bar{a}^i_{[x]+t:\overline{r}|}$, where x is the age at which disablement took place, t is the duration of the disability and r is the unexpired term of the disability benefit. In practice, full 'select' tables, in which the mortality of disabled lives is represented as a function of the duration as well as the age, are not usually constructed, but 'select' annuities (i.e. annuities applicable to lives who have just become disabled) and 'ultimate' annuities (i.e. annuities applicable to lives whose disability has lasted for more than a certain number of years) may be available. It will probably be considered sufficient to value by the 'ultimate' annuity $\bar{a}^i_{y:\overline{r}|}$, y being the age attained. As the selection which takes place on disablement is of the 'reversed' or 'negative' type, this errs on the safe side.

*31·9. The Hamza method

If it can be assumed that the mortality of disabled lives is a function of age only and is not affected by the duration since disablement an expression for $\bar{a}_x^{ai}$ can be obtained as follows:

Using the American notation mentioned in section 2 of this chapter, we have

$$l_x\bar{a}_x = l_x^{aa}\bar{a}_x^{aa} + l_x^{aa}\bar{a}_x^{ai} + l_x^{ii}\bar{a}_x^{i}.$$

$$\therefore \quad \bar{a}_x^{ai} = \frac{l_x}{l_x^{aa}}\bar{a}_x - \frac{l_x^{ii}}{l_x^{aa}}\bar{a}_x^{i} - \bar{a}_x^{aa}. \qquad (31\cdot6)$$

On the basis of this formula $\bar{a}_x^{ai}$ can be evaluated by means of the commutation functions N_x^{aa}, N_x^{ii}, etc., calculated from a combined mortality and disability table. The method is known as the Hamza method because it was originally developed by Hamza in a paper submitted at the Third International Congress of Actuaries. Because of the inaccuracy of the underlying assumption the method is now rarely employed.

31·10. Conclusion

In this chapter, only a broad and elementary treatment of the subject of disability benefits has been attempted. The limited extent to which the business is transacted in Great Britain does not justify more detailed consideration here. In the United States and Canada, there has been a very wide demand for this type of insurance and elaborate formulae have been devised for dealing with different types of benefit, with variations in the waiting period, etc. The reader who wishes to pursue the matter is referred to *Actuarial Studies No.* 5: *Disability benefits in life assurance policies* (second edition) by A. Hunter and J. T. Phillips, published in 1932 by the Actuarial Society of America.

EXAMPLES

Example 31·1

(*a*) A life office issues whole-life non-profit assurances, payable at the moment of death, at yearly premiums payable throughout life calculated on a certain mortality table at 3 per cent interest. Show how you would calculate on a net basis what extra yearly premium per £100 assured would be required in the case of a life aged 30, to provide that, in the event of permanent disablement occurring prior to age 60, premiums shall cease and a lump sum equal to 10 per cent of the sum assured shall be paid immediately. The full sum assured is to be payable on death whether or not the disability benefit has been paid. Assume that the mortality both before and after disablement is that assumed in the premium calculation and that the force of disablement will be constant and equal to ·009662 over the range of ages concerned.

(*b*) If the office premium is 1·15 times the net premium for the whole-life assurance and a loading of 15 per cent for expenses is required on the disability portion of the premium, show how you would calculate the extra premium on the assumption that there is no reduction in the expenses of the life policy when premiums are waived during disability.

Solution

(*a*) Let accented functions relate to survival without disability. Then the force of decrement of $v^t_{\cdot 03}\cdot {}_tp'_{30}$ is

$$\mu_{30+t} + \cdot 009662 + \delta_{\cdot 03} = \mu_{30+t} + \delta_{\cdot 04}.$$

$$\therefore \quad v^t_{\cdot 03}\cdot {}_tp'_{30} = v^t_{\cdot 04}\cdot {}_tp_{30}.$$

The value of the benefits is

$$100\bar{A}_{30}^{\cdot 03}+10\int_0^{30} v_{\cdot 03}^{\prime t}\cdot {}_tp_{30}^{\prime}\times \cdot 009662\,dt=100\bar{A}_{30}^{\cdot 03}+\cdot 09662\bar{a}_{30:\overline{30}|}^{\cdot 04}.$$

The annuity for valuing the premiums is

$$\ddot{a}_{30}^{\prime}+v_{\cdot 03}^{\prime 30}\cdot {}_{30}p_{30}^{\prime}(\ddot{a}_{60}-\ddot{a}_{60}^{\prime})=\ddot{a}_{30}^{\cdot 04}+v_{\cdot 04}^{30}\cdot {}_{30}p_{30}(\ddot{a}_{60}^{\cdot 03}-\ddot{a}_{60}^{\cdot 04}).$$

The net annual premium is therefore

$$\frac{100\bar{A}_{30}^{\cdot 03}+\cdot 09662\bar{a}_{30:\overline{30}|}^{\cdot 04}}{\ddot{a}_{30}^{\cdot 04}+v_{\cdot 04}^{30}\cdot {}_{30}p_{30}(\ddot{a}_{60}^{\cdot 03}-\ddot{a}_{60}^{\cdot 04})}$$

and the net extra annual premium is the difference between this amount and $1\cdot 015\times 100P_{30}^{\cdot 03}$.

(*b*) Let the annuity for valuing the premiums, set out above, be represented by $\ddot{a}''$. Then the net value of the 'waiver of premium' portion of the benefit is obtained by multiplying the office premium for the life assurance policy by $(\ddot{a}_{30}^{\cdot 03}-\ddot{a}'')$.

The office extra annual premium is therefore

$$\frac{1\cdot 15\{\cdot 09662\bar{a}_{30:\overline{30}|}^{\cdot 04}+1\cdot 015\times 115P_{30}^{\cdot 03}(\ddot{a}_{30}^{\cdot 03}-\ddot{a}'')\}}{\ddot{a}''},$$

where $$\ddot{a}''=\ddot{a}_{30}^{\cdot 04}+v_{\cdot 04}^{30}\cdot {}_{30}p_{30}(\ddot{a}_{60}^{\cdot 03}-\ddot{a}_{60}^{\cdot 04}).$$

Example 31·2

(*a*) On the basis of the multiple-decrement table reproduced on page 247 of this book (assuming that the figures in the column headed i_x represent persons becoming totally and permanently disabled) calculate the net annual premium ceasing at age 60 to provide, in respect of a life aged 30, a disability benefit consisting of a waiver of premiums and an income benefit of 10 per cent of the sum assured, the basic policy being a whole-life assurance for £1,000 with premiums payable throughout life and the annual premium being £15. The full sum assured is to be payable on death whether or not the disability benefit has been paid. Disablement occurring after age 60 does not quality for benefit. Rate of interest—3 per cent.

You should assume that the value of a continuous annuity of 1 to a life becoming disabled between ages 30 and 60 is 10·0.

(*b*) If the policy is effected on 30 June 1956 and the life assured attains age 30 on that date, what reserve should be held on 31 December 1966 in respect of the disability benefit, disablement not having occurred?

Note

The multiple-decrement table at the end of this book has been introduced into this question merely in order to illustrate the practical application of the formulae developed in the chapter. A pension fund table would not normally be used for valuing disability benefits.

Solution

(*a*) The net annual premium may be expressed as

$$115\frac{M_{30}^{ia}-M_{60}^{ia}}{N_{30}-N_{60}}=115\frac{M_{30}^{ia}-M_{60}^{ia}}{\frac{1}{2}D_{30}+\bar{N}_{30}-\bar{N}_{60}-\frac{1}{2}D_{60}}.$$

Since the values of M_x^{ia} shown at the end of this book are based on an annuity value of 11·00 we must multiply them by 10/11. The net annual premium is therefore

$$115\times\frac{10}{11}\cdot\frac{25{,}032-6{,}052}{17{,}125+673{,}213-31{,}687-5{,}362}$$

$$=\frac{115\times 10\times 18{,}980}{11\times 653{,}289}=3{\cdot}037.$$

(*b*) The reserve on 31 December 1966 is

$$\frac{115(M_{40\frac{1}{2}}^{ia}-M_{60}^{ia})-3{\cdot}037(N_{41}-N_{60})}{D_{40\frac{1}{2}}}.$$

$$115\times\tfrac{10}{11}(23{,}333-160-6{,}052)=1{,}790\times 10^3$$

$$3{\cdot}037(11{,}970+355{,}952-31{,}687-5{,}362)=1{,}005\times 10^3$$

$$785$$

$$\tfrac{1}{2}(24{,}758+23{,}939)=24{\cdot}348\times 10^3.$$

The reserve is therefore $\dfrac{785}{24{\cdot}348}=£32.$

Rough check

The method, mentioned in the chapter, of reserving half a year's disability premium is based on the assumption that $v^{\frac{1}{2}}(aq)^i_x \bar{a}^i_{x+\frac{1}{2}}$ is approximately constant for all values of x. In the present instance this assumption is not fulfilled. From the table on page 246, since $q^i_x \doteq (aq)^i_x$ and $\bar{a}^i_{x+\frac{1}{2}} = 10$, the average value of $v^{\frac{1}{2}}(aq)^i_x \bar{a}^i_{x+\frac{1}{2}}$ when $30 \leqslant x < 40$ is ·0054. The reserve is therefore roughly

$$10\frac{D_{35}}{D_{40}}(3{\cdot}037 - 115 \times {\cdot}0054) + \frac{3{\cdot}037}{2} \doteq £30.$$

This checks the arithmetic of (*a*) as well as (*b*).

INTERNATIONAL ACTUARIAL NOTATION

An international actuarial notation was adopted unanimously by the Second International Actuarial Congress held in London in May 1898. The following notation, which consists of the 1898 notation with a few alterations which had received unanimous or substantial support at the meetings of an International Committee appointed in 1937, was introduced by the Institute of Actuaries and the Faculty of Actuaries from the beginning of 1950; it was adopted internationally at the Madrid Congress in 1954.

The general principles on which the system is based are as follows:

To each fundamental symbolic letter are attached signs and letters each having its own signification.

The lower space to the left is reserved for signs indicating the conditions relative to the duration of the operations and to their position with regard to time.

The lower space to the right is reserved for signs indicating the conditions relative to ages and the order of succession of the events.

The upper space to the right is reserved for signs indicating the periodicity of the events.

The upper space to the left is free, and in it can be placed signs corresponding to other notions.

In what follows these two conventions are used:

A letter enclosed in brackets, thus (x), denotes 'a person aged x'. A letter or number enclosed in a right angle, thus $\overline{n|}$ or $\overline{15|}$, denotes a term-certain of years.

FUNDAMENTAL SYMBOLIC LETTERS

Interest

i = the effective rate of interest, namely, the total interest earned on 1 in a year on the assumption that the actual interest (if receivable otherwise than yearly) is invested forthwith as it becomes due on the same terms as the original principal.

$v=(1+i)^{-1}$ = the present value of 1 due a year hence.

$d=1-v$ = the discount on 1 due a year hence.

$\delta=\log_e(1+i)=-\log_e(1-d)$ = the force of interest or the force of discount.

Mortality Tables

l = number living.

d = number dying.

p = probability of living.

q = probability of dying.

μ = force of mortality.

m = central death rate.

a = present value of an annuity.

s = amount of an annuity.

e = expectation of life.

A = present value of an assurance.

E = present value of an endowment.

$\left.\begin{matrix}P\\ \pi\end{matrix}\right\}$ = premium per annum. P generally refers to net premiums, π to special premiums.

V = policy value.

W = paid-up policy.

The methods of using the foregoing principal letters and their precise meaning when added to by suffixes, etc., follow.

Interest

$i^{(m)}=m\{(1+i)^{1/m}-1\}$ = the nominal rate of interest, convertible m times a year.

$a_{\overline{n}|}=v+v^2+\ldots+v^n$ = the value of an annuity-certain of 1 per annum for n years, the payments being made at the end of each year.

$\ddot{a}_{\overline{n}|}=1+v+v^2+\ldots+v^{n-1}$ = the value of a similar annuity, the payments being made at the beginning of each year.

$s_{\overline{n}|}=1+(1+i)+(1+i)^2+\ldots+(1+i)^{n-1}$ = the amount of an annuity-certain of 1 per annum for n years, the payments being made at the end of each year.

$\ddot{s}_{\overline{n}|}=(1+i)+(1+i)^2+\ldots+(1+i)^n$ = the amount of a similar annuity, the payments being made at the beginning of each year.

The diaeresis or trema above the letters *a* and *s* is used as a symbol of acceleration of payments.

Mortality Tables

The ages of the lives involved are denoted by letters placed as suffixes in the lower space to the right. Thus:

l_x = the number of persons who attain age x according to the mortality table.

$d_x = l_x - l_{x+1}$ = the number of persons who die between ages x and $x+1$ according to the mortality table.

p_x = the probability that (x) will live 1 year.

q_x = the probability that (x) will die within 1 year.

$\mu_x = -\frac{1}{l_x}\frac{dl_x}{dx}$ = the force of mortality at age x.

m_x = the central death-rate for the year of age x to $x+1$

$$= d_x \Big/ \int_0^1 l_{x+t}\,dt.$$

e_x = the curtate 'expectation of life' (or average after-lifetime) of (x).

In the following it is always to be understood (unless otherwise expressed) that the annual payment of an annuity is 1, that the sum assured in any case is 1, and that the symbols indicate the present values:

a_x = an annuity, first payment at the end of a year, to continue during the life of (x).

$\ddot{a}_x = 1 + a_x$ = an 'annuity-due' to continue during the life of (x), the first payment to be made at once.

A_x = an assurance payable at the end of the year of death of (x).

Note. $e_x = a_x$ at rate of interest $i = 0$.

A letter or number at the lower left corner of the principal symbol denotes the number of years involved in the probability or benefit in question. Thus:

${}_np_x$ = the probability that (x) will live n years.

${}_nq_x$ = the probability that (x) will die within n years.

Note. When $n=1$ it is customary to omit it, as shown on page 221, provided no ambiguity is introduced.

${}_nE_x = v^n {}_np_x$ = the value of an endowment on (x) payable at the end of n years if (x) be then alive.

If the letter or number comes before a perpendicular bar it shows that a period of deferment is meant. Thus:

${}_{n|}q_x$ = the probability that (x) will die in a year, deferred n years; that is, that he will die in the $(n+1)$th year.

${}_{n|}a_x$ = an annuity on (x) deferred n years; that is, that the first payment is to be made at the end of $(n+1)$ years.

${}_{n|t}a_x$ = an intercepted or deferred temporary annuity on (x) deferred n years and, after that, to run for t years.

A letter or number in brackets at the upper right corner of the principal symbol shows the number of intervals into which the year is to be divided. Thus:

$a_x^{(m)}$ = an annuity on (x) payable by m instalments of $1/m$ each throughout the year, the first payment being one of $1/m$ at the end of the first $1/m$th of a year.

$\ddot{a}_x^{(m)}$ = a similar annuity but the first payment of $1/m$ is to be made at once, so that

$\ddot{a}_x^{(m)} = 1/m + a_x^{(m)}$.

$A_x^{(m)}$ = an assurance payable at the end of that fraction $1/m$ of a year in which (x) dies.

If $m \to \infty$, then instead of writing (∞) a bar is placed over the principal symbol. Thus:

$\bar{a}$ = a continuous or momently annuity.

$\bar{A}$ = an assurance payable at the moment of death.

A small circle placed over the principal symbol shows that the benefit is to be complete. Thus:

$\mathring{a}$ = a complete annuity.

$\mathring{e}$ = the complete expectation of life.

Note. Some consider that $\bar{e}$ would be as appropriate as $\mathring{e}$. As $e_x = a_x$ at rate of interest $i=0$, so also the complete expectation of life $= \bar{a}_x$ at rate of interest $i=0$.

When more than one life is involved the following rules are observed:

If there are two or more letters or numbers in a suffix without any distinguishing mark, joint lives are intended. Thus:

$$l_{xy}=l_x \times l_y, \quad d_{xy}=l_{xy}-l_{x+1:y+1}.$$

Note. When, for the sake of distinctness, it is desired to separate letters or numbers in a suffix, a colon is placed between them. A colon is used instead of a point or comma to avoid confusion with decimals when numbers are involved.

a_{xyz} = an annuity, first payment at the end of a year, to continue during the joint lives of (x), (y) and (z).

A_{xyz} = an assurance payable at the end of the year of the failure of the joint lives (x), (y) and (z).

In place of a life a term-certain may be involved. Thus:

$a_{x:\overline{n}|}$ = an annuity to continue during the joint duration of the life of (x) and a term of n years certain; that is, a temporary annuity for n years on the life of (x).

$A_{x:\overline{n}|}$ = an assurance payable at the end of the year of death of (x) if he die within n years, or at the end of n years if (x) be then alive; that is, an endowment assurance for n years.

If a perpendicular bar separates the letters in the suffix, then the status after the bar is to follow the status before the bar. Thus:

$a_{y|x}$ = a reversionary annuity, that is, an annuity on the life of (x) after the death of (y).

$A_{z|xy}$ = an assurance payable on the failure of the joint lives (x) and (y) provided both these lives survive (z).

If a horizontal bar appears above the suffix then survivors of the lives, and not joint lives, are intended. The number of survivors can be denoted by a letter or number over the right end of the bar. If that letter, say r, is not distinguished by any mark, then the meaning is *at least* r survivors; but if it is enclosed in square brackets, $[r]$, then the meaning is *exactly* r survivors. If no letter or number appears over the bar, then unity is supposed and the meaning is *at least one* survivor. Thus:

$a_{\overline{xyz}}$ = an annuity payable so long as at least one of the three lives (x), (y) and (z) is alive.

$a_{\overline{xyz}}^{\;2}$ = an annuity payable so long as at least two of the three lives (x), (y) and (z) are alive.

$p_{xyz}^{[2]}$ = probability that exactly two of the three lives (x), (y) and (z) will survive a year.

${}_nq_{\overline{xy}}$ = probability that the survivor of the two lives (x) and (y) will die within n years $= {}_nq_x \times {}_nq_y$.

${}_nA_{\overline{xy}}$ = an assurance payable at the end of the year of death of the survivor of the lives (x) and (y) provided the death occurs within n years.

When numerals are placed above or below the letters of the suffix, they designate the order in which the lives are to fail. The numeral placed *over* the suffix points out the life whose failure will finally determine the event; and the numerals placed *under* the suffix indicate the order in which the other lives involved are to fail. Thus:

$A_{\overset{1}{x}y}$ = an assurance payable at the end of the year of death of (x) if he dies first of the two lives (x) and (y).

$A_{\overset{2}{x}yz}$ = an assurance payable at the end of the year of death of (x) if he dies second of the three lives (x), (y) and (z).

$A_{\overset{2}{x}\underset{1}{y}z}$ = an assurance payable at the end of the year of death of (x) if he dies second of the three lives, (y) having died first.

$A_{\overline{xy}:\underset{3}{z}}$ = an assurance payable at the end of the year of death of the survivor of (x) and (y) if he dies before (z).

$A_{\overset{1}{x}:\overline{n}|}$ = an assurance payable at the end of the year of death of (x) if he dies within a term of n years.

$\left.\begin{matrix} a_{\overline{y\underset{1}{z}}|x} \\ \text{or} \\ a_{\overline{\overset{2}{y}z}|x} \end{matrix}\right\}$ = an annuity to (x) after the failure of the survivor of (y) and (z), provided (z) fails before (y).

Note. Sometimes to make quite clear that a joint-life status is involved a symbol ⌐¬ is placed above the lives included. Thus $A_{\overset{1}{\overline{xy}}:\overline{n}|}$ = a joint-life temporary assurance on (x) and (y).

In the case of reversionary annuities, distinction has sometimes to be made between those where the times of year at which payments are to take place are determined at the outset and those where the times depend on the failure of the preceding status. Thus:

$a_{y|x}$ = annuity to (x), first payment at the end of the year of the death of (y) or, on the average, about 6 months after his death.

$\hat{a}_{y|x}$ = annuity to (x), first payment 1 year after the death of (y).

$\mathring{\hat{a}}_{y|x}$ = complete annuity to (x), first payment 1 year after the death of (y).

ANNUAL PREMIUMS

The symbol P with the appropriate suffix or suffixes is used in simple cases, where no misunderstanding can occur, to denote the annual premium for a benefit. Thus:

P_x = the annual premium for an assurance payable at the end of the year of death of (x).

$P_{x:\overline{n}|}$ = the annual premium for an endowment assurance on (x) payable after n years or at the end of the year of death of (x) if he die within n years.

P^{1}_{xy} = the annual premium for a contingent assurance payable at the end of the year of death of (x) if he die before (y).

In all these cases it is optional to use the symbol P in conjunction with the principal symbol denoting the benefit. Thus instead of $P_{x:\overline{n}|}$ we may write $P(A_{x:\overline{n}|})$. In the more complicated cases it is necessary to use the two symbols in this way. Suffixes, etc., showing the conditions of the benefit are to be attached to the principal letter, and those showing the condition of payment of the premium are to be attached to the subsidiary symbol P. Thus:

${}_nP(\bar{A}_x)$ = the annual premium payable for n years only for an assurance payable at the moment of the death of (x).

$P_{xy}(A_x)$ = the annual premium payable during the joint lives of (x) and (y) for an assurance payable at the end of the year of death of (x).

${}_nP({}_n|a_x)$ = the annual premium payable for n years only for an annuity on (x) deferred n years.

${}_tP^{(m)}(A_{x:\overline{n}|})$ = the annual premium payable for t years only, by m instalments throughout the year, for an endowment assurance for n years on (x) (see below as to $P^{(m)}$).

Notes. (1) As a general rule the symbol P could be used without the principal symbol in the case of assurances where the sum assured is payable at the end of the year of death, but if it is payable at other times, or if the benefit is an annuity, then the principal symbol should be used.

(2) $P_x^{(m)}$. A point which was not brought out when the international system was adopted is that there are two kinds of premiums payable m times a year, viz. those which cease on payment of the instalment immediately preceding death and those which continue to be payable to the end of the year of death. To distinguish the latter the m is sometimes enclosed in square brackets, thus $P^{[m]}$.

POLICY VALUES AND PAID-UP POLICIES

${}_tV_x$ = the value of an ordinary whole-life assurance on (x) which has been t years in force, the premium then just due being unpaid.

${}_tW_x$ = the paid-up policy the present value of which is ${}_tV_x$.

The symbols V and W may, in simple cases, be used alone, but in the more complicated cases it is necessary to insert the full symbol for the benefit thus:

$$ {}_tV^{(m)}(\bar{A}_{x:\overline{n}|}) \text{ (corresponding to } P^{(m)}(\bar{A}_{x:\overline{n}|})), \quad {}_tV({}_{n|}a_x). $$

Note. As a general rule V or W can be used as the main symbol if the sum assured is payable at the end of the year of death and the premium is payable periodically throughout the duration of the assurance. If the premium is payable for a limited number of years, say n, the policy value after t years could be written ${}_tV[{}_nP(A)]$, or, if desired, ${}^n_tV(A)$.

In investigations where modified premiums and policy values are in question such modification may be denoted by adding accents to the symbols. Thus, when a premium other than the net premium (a valuation premium) is used in a valuation it may be denoted by P' and the corresponding policy value by V'. Similarly, the office (or commercial) premium may be denoted by P'' and the corresponding paid-up policy by W''.

COMPOUND SYMBOLS

(Ia) = an annuity, (IA) = an assurance } commencing at 1 and increasing 1 per annum.

If the whole benefit is to be temporary the symbol of limitation is placed outside the brackets. Thus:

$(Ia)_{x:\overline{n}|}$ = a temporary increasing annuity.
$(IA)^{1}_{x:\overline{n}|}$ = a temporary increasing assurance.

If only the increase is to be temporary but the benefit is to continue thereafter, then the symbol of limitation is placed immediately after the symbol I. Thus:

$(I_{\overline{n}|}a)_x$ = a whole-life annuity, $(I_{\overline{n}|}A)_x$ = a whole-life assurance } increasing for n years and thereafter stationary.

If the benefit is a decreasing one, the corresponding symbol is D. From the nature of the case this decrease must have a limit, as otherwise negative values might be implied. Thus:

$(D_{\overline{n}|}A)^{1}_{x:\overline{n}|}$ = a temporary assurance commencing at n and decreasing by 1 in each successive year.

If the benefit is a varying one the corresponding symbol is v. Thus:

(va) = a varying annuity.

COMMUTATION COLUMNS

Single lives

$$D_x = v^x l_x,$$
$$N_x = D_x + D_{x+1} + D_{x+2} + \text{etc.},$$
$$S_x = N_x + N_{x+1} + N_{x+2} + \text{etc.},$$
$$C_x = v^{x+1} d_x,$$
$$M_x = C_x + C_{x+1} + C_{x+2} + \text{etc.},$$
$$R_x = M_x + M_{x+1} + M_{x+2} + \text{etc.}$$

When it is desired to construct the assurance columns so as to give directly assurances payable at the moment of death the symbols are distinguished by a bar placed over them. Thus:

$$\bar{C}_x = v^{x+\frac{1}{2}} d_x \text{ which is an approximation to } \int_0^1 v^{x+t} \mu_{x+t} l_{x+t} dt,$$

$$\bar{M}_x = \bar{C}_x + \bar{C}_{x+1} + \bar{C}_{x+2} + \text{etc.},$$

$$\bar{R}_x = \bar{M}_x + \bar{M}_{x+1} + \bar{M}_{x+2} + \text{etc.}$$

Joint lives

$$D_{xy} = v^{\frac{1}{2}(x+y)} l_{xy},$$

$$N_{xy} = D_{xy} + D_{x+1\,:\,y+1} + D_{x+2\,:\,y+2} + \text{etc.},$$

$$C_{xy} = v^{\frac{1}{2}(x+y)+1} d_{xy},$$

$$M_{xy} = C_{xy} + C_{x+1\,:\,y+1} + C_{x+2\,:\,y+2} + \text{etc.},$$

$$C^{1}_{xy} = v^{\frac{1}{2}(x+y)+1} d_x l_{y+\frac{1}{2}},$$

$$M^{1}_{xy} = C^{1}_{xy} + C^{\;1}_{x+1\,:\,y+1} + C^{\;1}_{x+2\,:\,y+2} + \text{etc.}$$

SELECTION

If the suffix to a symbol which denotes the age is enclosed in a square bracket it indicates the age at which the life was selected. To this may be added, outside the bracket, the number of years which have elapsed since selection, so that the total suffix denotes the present age. Thus:

$l_{[x]+t}$ = the number in the select life table who were selected at age x and have attained age $x+t$.

$d_{[x]+t} = l_{[x]+t} - l_{[x]+t+1}$.

$a_{[x]}$ = value of an annuity on a life now aged x and now select.

$a_{[x-n]+n}$ = value of an annuity on a life now aged x and select n years ago at age $x-n$.

$$N_{[x]} = D_{[x]} + D_{[x]+1} + D_{[x]+2} + \ldots.$$

$$\ddot{a}_{[x]} = N_{[x]} \div D_{[x]} = 1 + a_{[x]},$$

and similarly for other functions.

TABLES

The following tables, which are extracted from *Actuarial Tables for Examination Purposes* (published for the Institute of Actuaries and the Faculty of Actuaries), are reproduced here for the convenience of students.

SECTION I

COMPOUND INTEREST TABLES

Rate of interest 2 per cent

2 per cent
COMPOUND INTEREST TABLES

n	$(1+i)^n$	v^n	$s_{\overline{n}\mid}$	$a_{\overline{n}\mid}$	$(a_{\overline{n}\mid})^{-1}$	n
1	1·020 00	·980 39	1·000 0	0·980 4	1·020 000	1
2	1·040 40	·961 17	2·020 0	1·941 6	0·515 050	2
3	1·061 21	·942 32	3·060 4	2·883 9	·346 755	3
4	1·082 43	·923 85	4·121 6	3·807 7	·262 624	4
5	1·104 08	·905 73	5·204 0	4·713 5	·212 158	5
6	1·126 16	·887 97	6·308 1	5·601 4	·178 526	6
7	1·148 69	·870 56	7·434 3	6·472 0	·154 512	7
8	1·171 66	·853 49	8·583 0	7·325 5	·136 510	8
9	1·195 09	·836 76	9·754 6	8·162 2	·122 515	9
10	1·218 99	·820 35	10·949 7	8·982 6	·111 327	10
11	1·243 37	·804 26	12·168 7	9·786 8	·102 178	11
12	1·268 24	·788 49	13·412 1	10·575 3	·094 560	12
13	1·293 61	·773 03	14·680 3	11·348 4	·088 118	13
14	1·319 48	·757 88	15·973 9	12·106 2	·082 602	14
15	1·345 87	·743 01	17·293 4	12·849 3	·077 825	15
16	1·372 79	·728 45	18·639 3	13·577 7	·073 650	16
17	1·400 24	·714 16	20·012 1	14·291 9	·069 970	17
18	1·428 25	·700 16	21·412 3	14·992 0	·066 702	18
19	1·456 81	·686 43	22·840 6	15·678 5	·063 782	19
20	1·485 95	·672 97	24·297 4	16·351 4	·061 157	20
21	1·515 67	·659 78	25·783 3	17·011 2	·058 785	21
22	1·545 98	·646 84	27·299 0	17·658 0	·056 631	22
23	1·576 90	·634 16	28·845 0	18·292 2	·054 668	23
24	1·608 44	·621 72	30·421 9	18·913 9	·052 871	24
25	1·640 61	·609 53	32·030 3	19·523 5	·051 220	25
26	1·673 42	·597 58	33·670 9	20·121 0	·049 699	26
27	1·706 89	·585 86	35·344 3	20·706 9	·048 293	27
28	1·741 02	·574 37	37·051 2	21·281 3	·046 990	28
29	1·775 84	·563 11	38·792 2	21·844 4	·045 778	29
30	1·811 36	·552 07	40·568 1	22·396 5	·044 650	30
31	1·847 59	·541 25	42·379 4	22·937 7	·043 596	31
32	1·884 54	·530 63	44·227 0	23·468 3	·042 611	32
33	1·922 23	·520 23	46·111 6	23·988 6	·041 687	33
34	1·960 68	·510 03	48·033 8	24·498 6	·040 819	34
35	1·999 89	·500 03	49·994 5	24·998 6	·040 002	35
36	2·039 89	·490 22	51·994 4	25·488 8	·039 233	36
37	2·080 69	·480 61	54·034 3	25·969 5	·038 507	37
38	2·122 30	·471 19	56·114 9	26·440 6	·037 821	38
39	2·164 74	·461 95	58·237 2	26·902 6	·037 171	39
40	2·208 04	·452 89	60·402 0	27·355 5	·036 556	40
41	2·252 20	·444 01	62·610 0	27·799 5	·035 972	41
42	2·297 24	·435 30	64·862 2	28·234 8	·035 417	42
43	2·343 19	·426 77	67·159 5	28·661 6	·034 890	43
44	2·390 05	·418 40	69·502 7	29·080 0	·034 388	44
45	2·437 85	·410 20	71·892 7	29·490 2	·033 910	45
46	2·486 61	·402 15	74·330 6	29·892 3	·033 453	46
47	2·536 34	·394 27	76·817 2	30·286 6	·033 018	47
48	2·587 07	·386 54	79·353 5	30·673 1	·032 602	48
49	2·638 81	·378 96	81·940 6	31·052 1	·032 204	49
50	2·691 59	·371 53	84·579 4	31·423 6	·031 823	50
60	3·281 03	·304 78	114·051 5	34·760 9	·028 768	60
70	3·999 56	·250 03	149·977 9	37·498 6	·026 668	70
80	4·875 44	·205 11	193·772 0	39·744 5	·025 161	80
90	5·943 13	·168 26	247·156 7	41·586 9	·024 046	90
100	7·244 65	·138 03	312·232 3	43·098 4	·023 203	100

SECTION II

ENGLISH LIFE TABLE NO. 10, MALES

Basis of the table

This table is based on the mortality of the male population of England and Wales in the years 1930 to 1932. The method of construction is explained in the Registrar-General's Decennial Supplement 1931, England and Wales, Part I (Life Tables).

Values at young ages

The value of μ_0 is indeterminate and therefore no value has been included in the table.

The value shown for $\mathring{e}_0$ (58·74) is based on the approximate formula $e_0+\frac{1}{2}$. An adjustment for the unequal distribution of deaths in the first year of life would reduce this value by ·02 only. Values of $\mathring{e}_x$ for other ages have been calculated by the formula

$$\mathring{e}_x=e_x+\tfrac{1}{2}-\tfrac{1}{12}\mu_x.$$

ENGLISH LIFE TABLE No. 10—MALES

Age x	l_x	d_x	p_x	q_x	μ_x	$\mathring{e}_x$	Age x
0	100 000	7 186	·928 14	·071 86		58·74	0
1	92 814	1 420	·984 70	·015 30	·031 46	62·25	1
2	91 394	600	·993 43	·006 57	·008 18	62·21	2
3	90 794	400	·995 59	·004 41	·004 82	61·62	3
4	90 394	325	·996 41	·003 59	·003 84	60·89	4
5	90 069	309	·996 57	·003 43	·003 52	60·11	5
6	89 760	233	·997 40	·002 60	·003 04	59·31	6
7	89 527	195	·997 82	·002 18	·002 35	58·47	7
8	89 332	165	·998 15	·001 85	·002 00	57·59	8
9	89 167	144	·998 39	·001 61	·001 72	56·70	9
10	89 023	130	·998 54	·001 46	·001 52	55·79	10
11	88 893	124	·998 61	·001 39	·001 41	54·87	11
12	88 769	125	·998 59	·001 41	·001 39	53·95	12
13	88 644	134	·998 49	·001 51	·001 45	53·02	13
14	88 510	150	·998 30	·001 70	·001 59	52·10	14
15	88 360	174	·998 03	·001 97	·001 82	51·19	15
16	88 186	200	·997 73	·002 27	·002 12	50·20	16
17	87 986	228	·997 41	·002 59	·002 44	49·40	17
18	87 758	249	·997 16	·002 84	·002 73	48·53	18
19	87 509	264	·996 98	·003 02	·002 94	47·66	19
20	87 245	276	·996 84	·003 16	·003 10	46·81	20
21	86 969	283	·996 75	·003 25	·003 22	45·95	21
22	86 686	286	·996 70	·003 30	·003 29	45·10	22
23	86 400	289	·996 66	·003 34	·003 33	44·25	23
24	86 111	287	·996 67	·003 33	·003 35	43·40	24
25	85 824	283	·996 70	·003 30	·003 32	42·54	25
26	85 541	280	·996 73	·003 27	·003 29	41·68	26
27	85 261	280	·996 72	·003 28	·003 28	40·82	27
28	84 981	281	·996 69	·003 31	·003 30	39·95	28
29	84 700	284	·996 65	·003 35	·003 33	39·08	29
30	84 416	287	·996 60	·003 40	·003 38	38·21	30
31	84 129	294	·996 51	·003 49	·003 45	37·34	31
32	83 835	303	·996 39	·003 61	·003 55	36·47	32
33	83 532	316	·996 22	·003 78	·003 70	35·60	33
34	83 216	331	·996 02	·003 98	·003 88	34·73	34
35	82 885	349	·995 79	·004 21	·004 10	33·87	35
36	82 536	369	·995 53	·004 47	·004 35	33·01	36
37	82 167	389	·995 26	·004 74	·004 61	32·15	37
38	81 778	411	·994 98	·005 02	·004 89	31·30	38
39	81 367	432	·994 69	·005 31	·005 18	30·46	39
40	80 935	455	·994 38	·005 62	·005 47	29·62	40
41	80 480	481	·994 02	·005 98	·005 81	28·78	41
42	79 999	511	·993 61	·006 39	·006 19	27·95	42
43	79 488	546	·993 13	·006 87	·006 64	27·13	43
44	78 942	585	·992 59	·007 41	·007 16	26·32	44
45	78 357	626	·992 01	·007 99	·007 72	25·51	45
46	77 731	669	·991 39	·008 61	·008 33	24·71	46
47	77 062	713	·990 75	·009 25	·008 97	23·92	47
48	76 349	756	·990 10	·009 90	·009 62	23·14	48
49	75 593	799	·989 43	·010 57	·010 28	22·36	49
50	74 794	844	·988 72	·011 28	·010 98	21·60	50
51	73 950	892	·987 94	·012 06	·011 73	20·84	51
52	73 058	946	·987 05	·012 95	·012 57	20·09	52
53	72 112	1 005	·986 07	·013 93	·013 52	19·34	53
54	71 107	1 066	·985 01	·014 99	·014 56	18·61	54

ENGLISH LIFE TABLE No. 10—MALES

Age x	l_x	d_x	p_x	q_x	μ_x	$\mathring{e}_x$	Age x
55	70 041	1130	·983 86	·016 14	·015 66	17·89	55
56	68 911	1202	·982 56	·017 44	·016 90	17·17	56
57	67 709	1280	·981 10	·018 90	·018 32	16·47	57
58	66 429	1362	·979 50	·020 50	·019 88	15·78	58
59	65 067	1447	·977 76	·022 24	·021 58	15·10	59
60	63 620	1536	·975 85	·024 15	·023 43	14·43	60
61	62 084	1633	·973 70	·026 30	·025 50	13·77	61
62	60 451	1738	·971 25	·028 75	·027 86	13·13	62
63	58 713	1849	·968 50	·031 50	·030 53	12·50	63
64	56 864	1965	·965 45	·034 55	·033 53	11·89	64
65	54 899	2081	·962 09	·037 91	·036 85	11·30	65
66	52 818	2198	·958 38	·041 62	·040 51	10·73	66
67	50 620	2312	·954 32	·045 68	·044 56	10·17	67
68	48 308	2422	·949 86	·050 14	·049 02	9·63	68
69	45 886	2525	·944 98	·055 02	·053 94	9·12	69
70	43 361	2617	·939 65	·060 35	·059 34	8·62	70
71	40 744	2695	·933 85	·066 15	·065 25	8·14	71
72	38 049	2757	·927 54	·072 46	·071 72	7·68	72
73	35 292	2801	·920 62	·079 38	·078 83	7·24	73
74	32 491	2826	·913 03	·086 97	·086 71	6·82	74
75	29 665	2824	·904 81	·095 19	·095 39	6·43	75
76	26 841	2791	·896 03	·103 97	·104 80	6·05	76
77	24 050	2724	·886 75	·113 25	·114 88	5·69	77
78	21 326	2626	·876 87	·123 13	·125 66	5·36	78
79	18 700	2501	·866 27	·133 73	·137 33	5·04	79
80	16 199	2349	·855 00	·145 00	·149 96	4·74	80
81	13 850	2173	·843 13	·156 87	·163 52	4·46	81
82	11 677	1976·6	·830 73	·169 27	·177 91	4·20	82
83	9 700·4	1768·3	·817 71	·182 29	·193 17	3·95	83
84	7 932·1	1555·2	·803 93	·196 07	·209 55	3·72	84
85	6 376·9	1342·2	·789 52	·210 48	·227 10	3·50	85
86	5 034·7	1135·0	·774 56	·225 44	·245 73	3·30	86
87	3 899·7	939·0	·759 22	·240 78	·265 41	3·12	87
88	2 960·7	755·6	·744 80	·255 20	·285 15	2·95	88
89	2 205·1	596·1	·729 69	·270 31	·304 69	2·79	89
90	1 609·0	460·4	·713 86	·286 14	·325 89	2·63	90
91	1 148·6	347·68	·697 30	·302 70	·348 55	2·49	91
92	800·92	256·28	·680 02	·319 98	·372 82	2·35	92
93	544·64	184·09	·662 00	·338 00	·398 78	2·22	93
94	360·55	128·62	·643 27	·356 73	·426 53	2·09	94
95	231·93	87·25	·623 81	·376 19	·456 19	1·97	95
96	144·68	57·342	·603 66	·396 34	·487 91	1·86	96
97	87·338	36·437	·582 80	·417 20	·521 80	1·76	97
98	50·901	22·330	·561 31	·438 69	·558 05	1·66	98
99	28·571	13·166	·539 19	·460 81	·596 72	1·57	99
100	15·405	7·448 3	·516 50	·483 50	·637 99	1·48	100
101	7·956 7	4·031 7	·493 29	·506 71	·681 94	1·40	101
102	3·925 0	2·081 8	·469 60	·530 40	·728 75	1·32	102
103	1·843 2	1·022 00	·445 53	·554 47	·778 52	1·25	103
104	·821 20	·475 35	·421 15	·578 85		1·18	104
105	·345 85						105

SECTION III

HYPOTHETICAL SELECT MORTALITY TABLE

ULTIMATE FUNCTIONS ONLY

The Hypothetical Select Mortality Table was constructed by R. E. Beard, M.B.E., F.I.A., and details are given in his paper entitled 'Some experiments in the use of the Incomplete Gamma Function for the Approximate Calculation of Actuarial Functions' submitted at the proceedings of the Institute of Actuaries' Centenary in 1948.

The table is designed to be mathematically smooth without introducing too much artificial constraint in the shape of the mortality curve. It exhibits mortality rates between those of the A 1924–29 Table and those of the A 1924–29 Light Table, but does not purport to represent any particular experience or to be suitable for any specific practical purpose.

HYPOTHETICAL SELECT MORTALITY TABLE

ULTIMATE VALUES

Age attained x	l_x	d_x	q_x	μ_x	$\log_{10}\mu_x$	e_x	$\log_{10} l_x$	Age attained x
15	500 000	907	·001 81	·001 81	$\bar{3}$·257 68	54·714	5·698 97	15
16	499 093	908	·001 82	·001 82	$\bar{3}$·260 07	53·814	5·698 18	16
17	498 185	908	·001 82	·001 82	$\bar{3}$·260 07	52·912	5·697 39	17
18	497 277	908	·001 83	·001 83	$\bar{3}$·262 45	52·008	5·696 60	18
19	496 369	910	·001 83	·001 83	$\bar{3}$·262 45	51·104	5·695 80	19
20	495 459	911	·001 84	·001 84	$\bar{3}$·264 82	50·197	5·695 01	20
21	494 548	915	·001 85	·001 85	$\bar{3}$·267 17	49·290	5·694 21	21
22	493 633	917	·001 86	·001 86	$\bar{3}$·269 51	48·381	5·693 40	22
23	492 716	921	·001 87	·001 87	$\bar{3}$·271 84	47·471	5·692 60	23
24	491 795	927	·001 88	·001 88	$\bar{3}$·274 16	46·560	5·691 78	24
25	490 868	933	·001 90	·001 89	$\bar{3}$·276 46	45·648	5·690 96	25
26	489 935	942	·001 92	·001 91	$\bar{3}$·281 03	44·735	5·690 14	26
27	488 993	950	·001 94	·001 93	$\bar{3}$·285 56	43·821	5·689 30	27
28	488 043	963	·001 97	·001 96	$\bar{3}$·292 26	42·906	5·688 46	28
29	487 080	977	·002 01	·001 99	$\bar{3}$·298 85	41·991	5·687 60	29
30	486 103	993	·002 04	·002 03	$\bar{3}$·307 50	41·076	5·686 73	30
31	485 110	1 014	·002 09	·002 07	$\bar{3}$·315 97	40·160	5·685 84	31
32	484 096	1 037	·002 14	·002 12	$\bar{3}$·326 34	39·244	5·684 93	32
33	483 059	1 065	·002 20	·002 17	$\bar{3}$·336 46	38·328	5·684 00	33
34	481 994	1 097	·002 28	·002 24	$\bar{3}$·350 25	37·413	5·683 04	34
35	480 897	1 136	·002 36	·002 32	$\bar{3}$·365 49	36·498	5·682 05	35
36	479 761	1 180	·002 46	·002 41	$\bar{3}$·382 02	35·585	5·681 02	36
37	478 581	1 233	·002 58	·002 52	$\bar{3}$·401 40	34·672	5·679 96	37
38	477 348	1 293	·002 71	·002 64	$\bar{3}$·421 60	33·762	5·678 84	38
39	476 055	1 362	·002 86	·002 78	$\bar{3}$·444 04	32·854	5·677 66	39
40	474 693	1 442	·003 04	·002 95	$\bar{3}$·469 82	31·948	5·676 41	40
41	473 251	1 535	·003 24	·003 14	$\bar{3}$·496 93	31·045	5·675 09	41
42	471 716	1 640	·003 48	·003 36	$\bar{3}$·526 34	30·146	5·673 68	42
43	470 076	1 761	·003 75	·003 61	$\bar{3}$·557 51	29·251	5·672 17	43
44	468 315	1 898	·004 05	·003 90	$\bar{3}$·591 06	28·361	5·670 54	44
45	466 417	2 054	·004 40	·004 23	$\bar{3}$·626 34	27·477	5·668 77	45
46	464 363	2 230	·004 80	·004 61	$\bar{3}$·663 70	26·598	5·666 86	46
47	462 133	2 429	·005 26	·005 03	$\bar{3}$·701 57	25·727	5·664 77	47
48	459 704	2 652	·005 77	·005 52	$\bar{3}$·741 94	24·863	5·662 48	48
49	457 052	2 901	·006 35	·006 06	$\bar{3}$·782 47	24·007	5·659 97	49
50	454 151	3 179	·007 00	·006 68	$\bar{3}$·824 78	23·160	5·657 20	50
51	450 972	3 488	·007 73	·007 38	$\bar{3}$·868 06	22·324	5·654 15	51
52	447 484	3 830	·008 56	·008 16	$\bar{3}$·911 69	21·498	5·650 78	52
53	443 654	4 205	·009 48	·009 04	$\bar{3}$·956 17	20·683	5·647 04	53
54	439 449	4 616	·010 50	·010 02	$\bar{2}$·000 87	19·881	5·642 91	54
55	434 833	5 066	·011 65	·011 12	$\bar{2}$·046 10	19·092	5·638 32	55
56	429 767	5 552	·012 92	·012 34	$\bar{2}$·091 32	18·317	5·633 23	56
57	424 215	6 076	·014 32	·013 69	$\bar{2}$·136 40	17·557	5·627 59	57
58	418 139	6 640	·015 88	·015 19	$\bar{2}$·181 56	16·812	5·621 32	58
59	411 499	7 238	·017 59	·016 85	$\bar{2}$·226 60	16·083	5·614 37	59
60	404 261	7 874	·019 48	·018 68	$\bar{2}$·271 38	15·371	5·606 66	60
61	396 387	8 539	·021 54	·020 69	$\bar{2}$·315 76	14·677	5·598 12	61
62	387 848	9 234	·023 81	·022 90	$\bar{2}$·359 84	14·000	5·588 66	62
63	378 614	9 950	·026 28	·025 33	$\bar{2}$·403 64	13·341	5·578 20	63
64	368 664	10 684	·028 98	·027 98	$\bar{2}$·446 85	12·701	5·566 63	64

HYPOTHETICAL SELECT MORTALITY TABLE

ULTIMATE VALUES

Age attained x	l_x	d_x	q_x	μ_x	$\log_{10}\mu_x$	e_x	$\log_{10}l_x$	Age attained x
65	357 980	11 426	·031 92	·030 88	$\bar{2}$·489 68	12·080	5·553 86	65
66	346 554	12 168	·035 11	·034 04	$\bar{2}$·531 99	11·479	5·539 77	66
67	334 386	12 898	·038 57	·037 49	$\bar{2}$·573 92	10·896	5·524 25	67
68	321 488	13 606	·042 32	·041 24	$\bar{2}$·615 32	10·333	5·507 16	68
69	307 882	14 277	·046 37	·045 31	$\bar{2}$·656 19	9·790	5·488 38	69
70	293 605	14 901	·050 75	·049 72	$\bar{2}$·696 53	9·266	5·467 76	70
71	278 704	15 460	·055 47	·054 51	$\bar{2}$·736 48	8·762	5·445 14	71
72	263 244	15 940	·060 55	·059 70	$\bar{2}$·775 97	8·276	5·420 36	72
73	247 304	16 328	·066 02	·065 31	$\bar{2}$·814 98	7·810	5·393 23	73
74	230 976	16 607	·071 90	·071 38	$\bar{2}$·853 58	7·362	5·363 57	74
75	214 369	16 766	·078 21	·077 94	$\bar{2}$·891 76	6·932	5·331 16	75
76	197 603	16 795	·084 99	·085 04	$\bar{2}$·929 62	6·520	5·295 79	76
77	180 808	16 681	·092 26	·092 71	$\bar{2}$·967 13	6·126	5·257 22	77
78	164 127	16 422	·100 06	·101 00	$\bar{1}$·004 32	5·748	5·215 18	78
79	147 705	16 014	·108 42	·109 97	$\bar{1}$·041 27	5·387	5·169 40	79
80	131 691	15 457	·117 37	·119 67	$\bar{1}$·077 99	5·042	5·119 56	80
81	116 234	14 759	·126 98	·130 18	$\bar{1}$·114 54	4·713	5·065 33	81
82	101 475	13 931	·137 29	·141 57	$\bar{1}$·150 97	4·398	5·006 36	82
83	87 544	12 985	·148 33	·153 93	$\bar{1}$·187 32	4·098	4·942 23	83
84	74 559	11 942	·160 17	·167 36	$\bar{1}$·223 65	3·812	4·872 50	84
85	62 617·0	10 826·5	·172 90	·181 98	$\bar{1}$·260 02	3·539	4·796 69	85
86	51 790·5	9 662·6	·186 57	·197 92	$\bar{1}$·296 49	3·279	4·714 25	86
87	42 127·9	8 479·2	·201 27	·215 34	$\bar{1}$·333 12	3·031	4·624 57	87
88	33 648·7	7 305·2	·217 10	·234 43	$\bar{1}$·370 01	2·795	4·526 97	88
89	26 343·5	6 168·6	·234 16	·255 41	$\bar{1}$·407 24	2·570	4·420 67	89
90	20 174·9	5 095·6	·252 57	·278 54	$\bar{1}$·444 89	2·356	4·304 81	90
91	15 079·3	4 108·6	·272 47	·304 13	$\bar{1}$·483 06	2·152	4·178 38	91
92	10 970·7	3 225·5	·294 01	·332 57	$\bar{1}$·521 88	1·958	4·040 23	92
93	7 745·2	2 458·1	·317 37	·364 33	$\bar{1}$·561 50	1·773	3·889 03	93
94	5 287·1	1 812·1	·342 74	·399 98	$\bar{1}$·602 04	1·597	3·723 22	94
95	$347\,500 \times 10^{-2}$	1 287·01	·370 36	·440 23	$\bar{1}$·643 68	1·430	3·540 95	95
96	$218\,799 \times 10^{-2}$	$87\,625 \times 10^{-2}$	·400 48	·486 00	$\bar{1}$·686 64	1·271	3·340 05	96
97	$131\,174 \times 10^{-2}$	$56\,849 \times 10^{-2}$	·433 39	·538 48	$\bar{1}$·731 17	1·120	3·117 85	97
98	$743\,251 \times 10^{-3}$	$34\,888 \times 10^{-2}$	·469 39	·599 20	$\bar{1}$·777 57	·977	2·871 13	98
99	$394\,374 \times 10^{-3}$	$20\,068 \times 10^{-2}$	·508 85	·670 20	$\bar{1}$·826 20	·841	2·595 90	99
100	$193\,695 \times 10^{-3}$	$10\,694 \times 10^{-2}$	·552 11	·754 29	$\bar{1}$·877 54	·713	2·287 11	100
101	$867\,541 \times 10^{-4}$	$52\,009 \times 10^{-3}$	·599 50	·855 37	$\bar{1}$·932 15	·592	1·938 29	101
102	$347\,448 \times 10^{-4}$	$22\,629 \times 10^{-3}$	·651 28	·979 10	$\bar{1}$·990 83	·479	1·540 88	102
103	$121\,162 \times 10^{-4}$	$85\,719 \times 10^{-4}$	·707 47	1·133 98	0·054 61	·373	1·083 37	103
104	$354\,431 \times 10^{-5}$	$27\,207 \times 10^{-4}$	·767 64	1·333 32	0·124 93	·276	0·549 53	104
105	$823\,539 \times 10^{-6}$	$68\,388 \times 10^{-5}$	·830 41	1·599 35	0·203 94	·189	$\bar{1}$·915 68	105
106	$139\,663 \times 10^{-6}$	$12\,466 \times 10^{-5}$	·892 58	1·972 07	0·294 92	·113	$\bar{1}$·145 08	106
107	$150\,029 \times 10^{-7}$	$14\,220 \times 10^{-6}$	·947 82	2·531 48	0·403 77	·053	$\bar{2}$·176 18	107
108	$782\,711 \times 10^{-9}$	$77\,179 \times 10^{-8}$	·986 04	3·464 24	0·539 61	·014	$\bar{4}$·893 60	108
109	$10\,923 \times 10^{-9}$	$10\,917 \times 10^{-9}$	·999 44	5·330 38	0·726 76	·000 6	$\bar{5}$·038 36	109
110	6×10^{-9}	6×10^{-9}	1·000 00	10·929 76	1·038 61	—	$\bar{9}$·784 04	110

HYPOTHETICAL SELECT MORTALITY TABLE

ULTIMATE VALUES **2 per cent**

Age attained x	D_x	N_x	S_x	C_x	M_x	R_x	Age attained x
15	371 507	12 345 436	296 445 944	660·7	129 440	6 532 765	15
16	363 562	11 973 929	284 100 508	648·5	128 779	6 403 325	16
17	355 785	11 610 367	272 126 579	635·7	128 131	6 274 546	17
18	348 173	11 254 582	260 516 212	623·3	127 495	6 146 415	18
19	340 723	10 906 409	249 261 630	612·4	126 872	6 018 920	19
20	333 430	10 565 686	238 355 221	601·1	126 259	5 892 048	20
21	326 291	10 232 256	227 789 535	591·9	125 658	5 765 789	21
22	319 301	9 905 965	217 557 279	581·5	125 066	5 640 131	22
23	312 459	9 586 664	207 651 314	572·6	124 485	5 515 065	23
24	305 759	9 274 205	198 064 650	565·0	123 912	5 390 580	24
25	299 199	8 968 446	188 790 445	557·5	123 347	5 266 668	25
26	292 775	8 669 247	179 821 999	551·9	122 790	5 143 321	26
27	286 482	8 376 472	171 152 752	545·7	122 238	5 020 531	27
28	280 320	8 089 990	162 776 280	542·3	121 692	4 898 293	28
29	274 281	7 809 670	154 686 290	539·4	121 150	4 776 601	29
30	268 363	7 535 389	146 876 620	537·5	120 610	4 655 451	30
31	262 564	7 267 026	139 341 231	538·1	120 073	4 534 841	31
32	256 877	7 004 462	132 074 205	539·5	119 535	4 414 768	32
33	251 301	6 747 585	125 069 743	543·2	118 995	4 295 233	33
34	245 830	6 496 284	118 322 158	548·5	118 452	4 176 238	34
35	240 462	6 250 454	111 825 874	556·9	117 904	4 057 786	35
36	235 190	6 009 992	105 575 420	567·1	117 347	3 939 882	36
37	230 011	5 774 802	99 565 428	581·0	116 780	3 822 535	37
38	224 920	5 544 791	93 790 626	597·3	116 199	3 705 755	38
39	219 913	5 319 871	88 245 835	616·8	115 601	3 589 556	39
40	214 984	5 099 958	82 925 964	640·3	114 985	3 473 955	40
41	210 128	4 884 974	77 826 006	668·2	114 344	3 358 970	41
42	205 340	4 674 846	72 941 032	699·9	113 676	3 244 626	42
43	200 614	4 469 506	68 266 186	736·8	112 976	3 130 950	43
44	195 943	4 268 892	63 796 680	778·6	112 239	3 017 974	44
45	191 323	4 072 949	59 527 788	826·0	111 461	2 905 735	45
46	186 745	3 881 626	55 454 839	879·2	110 635	2 794 274	46
47	182 204	3 694 881	51 573 213	938·9	109 756	2 683 639	47
48	177 693	3 512 677	47 878 332	1 005·0	108 817	2 573 883	48
49	173 204	3 334 984	44 365 655	1 077·8	107 812	2 465 066	49
50	168 730	3 161 780	41 030 671	1 157·9	106 734	2 357 254	50
51	164 263	2 993 050	37 868 891	1 245·6	105 576	2 250 520	51
52	159 797	2 828 787	34 875 841	1 340·9	104 330	2 144 944	52
53	155 323	2 668 990	32 047 054	1 443·3	102 989	2 040 614	53
54	150 834	2 513 667	29 378 064	1 553·3	101 546	1 937 625	54
55	146 323	2 362 833	26 864 397	1 671·3	99 993	1 836 079	55
56	141 783	2 216 510	24 501 564	1 795·7	98 322	1 736 086	56
57	137 207	2 074 727	22 285 054	1 926·7	96 526	1 637 764	57
58	132 590	1 937 520	20 210 327	2 064·2	94 599	1 541 238	58
59	127 926	1 804 930	18 272 807	2 206·0	92 535	1 446 639	59
60	123 211	1 677 004	16 467 877	2 352·8	90 329	1 354 104	60
61	118 443	1 553 793	14 790 873	2 501·5	87 976	1 263 775	61
62	113 619	1 435 350	13 237 080	2 652·0	85 475	1 175 799	62
63	108 739	1 321 731	11 801 730	2 801·6	82 823	1 090 324	63
64	103 805	1 212 992	10 479 999	2 949·3	80 021	1 007 501	64

HYPOTHETICAL SELECT MORTALITY TABLE

ULTIMATE VALUES 2 per cent

Age attained x	D_x	N_x	S_x	C_x	M_x	R_x	Age attained x
65	98 821	1 109 187	9 267 007	3 092·3	77 072	927 480	65
66	93 791	1 010 366	8 157 820	3 228·5	73 980	850 408	66
67	88 723	916 575	7 147 454	3 355·1	70 751	776 428	67
68	83 628	827 852	6 230 879	3 469·9	67 396	705 677	68
69	78 519	744 224	5 403 027	3 569·7	63 926	638 281	69
70	73 409	665 705	4 658 803	3 652·6	60 356	574 355	70
71	68 317	592 296	3 993 098	3 715·3	56 704	513 999	71
72	63 263	523 979	3 400 802	3 755·6	52 988	457 295	72
73	58 267	460 716	2 876 823	3 771·6	49 233	404 307	73
74	53 352	402 449	2 416 107	3 760·8	45 461	355 074	74
75	48 546	349 097	2 013 658	3 722·3	41 700	309 613	75
76	43 871	300 551	1 664 561	3 655·7	37 978	267 913	76
77	39 355	256 680	1 364 010	3 559·7	34 322	229 935	77
78	35 024	217 325	1 107 330	3 435·7	30 763	195 613	78
79	30 902	182 301	890 005	3 284·6	27 327	164 850	79
80	27 011	151 399	707 704	3 108·2	24 042	137 523	80
81	23 373	124 388	556 305	2 909·7	20 934	113 481	81
82	20 005	101 015	431 917	2 692·6	18 025	92 547	82
83	16 920	81 010	330 902	2 460·5	15 332	74 522	83
84	14 128	64 090	249 892	2 218·5	12 871	59 190	84
85	11 633·0	49 962·0	185 802·0	1 971·8	10 653·0	46 319·0	85
86	9 432·7	38 329·0	135 840·0	1 725·3	8 681·2	35 666·0	86
87	7 522·4	28 896·0	97 511·0	1 484·4	6 955·9	26 985·0	87
88	5 890·5	21 374·0	68 615·0	1 253·9	5 471·5	20 029·0	88
89	4 521·3	15 483·0	47 241·0	1 037·9	4 217·6	14 558·0	89
90	3 394·6	10 962·0	31 758·0	840·58	3 179·7	10 340·0	90
91	2 487·5	7 567·8	20 796·0	664·48	2 339·1	7 159·9	91
92	1 774·3	5 080·3	13 228·0	511·42	1 674·6	4 820·8	92
93	1 228·0	3 306·0	8 147·9	382·11	1 163·2	3 146·2	93
94	821·87	2 078·0	4 841·9	276·16	781·11	1 983·0	94
95	529·59	1 256·1	2 763·9	192·29	504·95	1 201·9	95
96	326·91	726·56	1 507·8	128·35	312·66	696·99	96
97	192·15	399·65	781·27	81·641	184·31	384·33	97
98	106·74	207·50	381·62	49·120	102·67	200·02	98
99	55·525	100·76	174·12	27·700	53·549	97·346	99
100	26·736	45·236	73·362	14·472	25·849	43·797	100
101	11·740	18·500	28·126	6·900 2	11·377	17·948	101
102	4·609 7	6·759 6	9·626 4	2·943 3	4·477 1	6·570 8	102
103	1·576 0	2·149 9	2·866 8	1·093 1	1·533 8	2·093 7	103
104	·451 97	·573 94	·716 90	·340 15	·440 72	·559 90	104
105	·102 96	·121 97	·142 96	·083 82	·100 57	·119 18	105
106	·017 12	·019 01	·020 99	·014 98	·016 75	·018 61	106
107	·001 80	·001 89	·001 98	·001 68	·001 77	·001 86	107
108	·000 09	·000 09	·000 09	·000 09	·000 09	·000 09	108
109	·000 00	·000 00	·000 00	·000 00	·000 00	·000 00	109

HYPOTHETICAL SELECT MORTALITY TABLE

ULTIMATE VALUES **2 per cent**

Age attained x	a_x	A_x	P_x	Age attained x	a_x	A_x	P_x
15	32·231	·348 42	·010 48	50	17·739	·632 57	·033 76
16	31·935	·354 22	·010 75	51	17·221	·642 73	·035 27
17	31·633	·360 14	·011 04	52	16·702	·652 89	·036 88
18	31·325	·366 18	·011 33	53	16·183	·663 07	·038 59
19	31·010	·372 36	·011 63	54	15·665	·673 23	·040 40
20	30·688	·378 67	·011 95	55	15·148	·683 37	·042 32
21	30·359	·385 11	·012 28	56	14·633	·693 47	·044 36
22	30·024	·391 69	·012 63	57	14·121	·703 51	·046 52
23	29·681	·398 40	·012 99	58	13·613	·713 47	·048 82
24	29·332	·405 26	·013 36	59	13·109	·723 35	·051 27
25	28·975	·412 26	·013 75	60	12·611	·733 12	·053 86
26	28·611	·419 40	·014 16	61	12·119	·742 77	·056 62
27	28·239	·426 69	·014 59	62	11·633	·752 29	·059 55
28	27·860	·434 12	·015 04	63	11·155	·761 67	·062 66
29	27·473	·441 70	·015 51	64	10·685	·770 88	·065 97
30	27·079	·449 43	·016 01	65	10·224	·779 91	·069 48
31	26·677	·457 31	·016 52	66	9·773	·788 77	·073 22
32	26·268	·465 34	·017 07	67	9·331	·797 44	·077 19
33	25·851	·473 52	·017 64	68	8·899	·805 90	·081 41
34	25·426	·481 85	·018 23	69	8·478	·814 15	·085 90
35	24·994	·490 32	·018 86	70	8·068	·822 19	·090 67
36	24·554	·498 94	·019 53	71	7·670	·830 01	·095 74
37	24·107	·507 71	·020 22	72	7·283	·837 59	·101 13
38	23·652	·516 62	·020 96	73	6·907	·844 95	·106 86
39	23·191	·525 67	·021 73	74	6·543	·852 10	·112 96
40	22·723	·534 85	·022 55	75	6·191	·858 99	·119 45
41	22·248	·544 16	·023 41	76	5·851	·865 68	·126 36
42	21·766	·553 60	·024 32	77	5·522	·872 12	·133 72
43	21·279	·563 15	·025 28	78	5·205	·878 31	·141 55
44	20·786	·572 82	·026 29	79	4·899	·884 31	·149 90
45	20·288	·582 58	·027 37	80	4·605	·890 10	·158 80
46	19·786	·592 44	·028 50	81	4·322	·895 66	·168 30
47	19·279	·602 38	·029 70	82	4·050	·901 00	·178 43
48	18·768	·612 39	·030 98	83	3·788	·906 14	·189 26
49	18·255	·622 46	·032 33	84	3·536	·911 06	·200 83

HYPOTHETICAL SELECT MORTALITY TABLE

JOINT LIFE ULTIMATE FUNCTIONS 2 per cent

x	D_{xx}	N_{xx}	a_{xx}
20	1 652 006	47 178 491	27·558
21	1 613 665	45 526 485	27·213
22	1 576 176	43 912 820	26·860
23	1 539 535	42 336 644	26·500
24	1 503 709	40 797 109	26·131
25	1 468 674	39 293 400	25·754
26	1 434 407	37 824 726	25·370
27	1 400 879	36 390 319	24·977
28	1 368 081	34 989 440	24·576
29	1 335 966	33 621 359	24·166
30	1 304 523	32 285 393	23·749
31	1 273 723	30 980 870	23·323
32	1 243 533	29 707 147	22·889
33	1 213 934	28 463 614	22·447
34	1 184 888	27 249 680	21·998
35	1 156 375	26 064 792	21·540
36	1 128 349	24 908 417	21·075
37	1 100 790	23 780 068	20·603
38	1 073 652	22 679 278	20·123
39	1 046 905	21 605 626	19·638
40	1 020 513	20 558 721	19·145
41	994 434	19 538 208	18·648
42	968 621	18 543 774	18·145
43	943 038	17 575 153	17·637
44	917 633	16 632 115	17·125
45	892 363	15 714 482	16·610
46	867 177	14 822 119	16·092
47	842 026	13 954 942	15·573
48	816 862	13 112 916	15·053
49	791 630	12 296 054	14·533
50	766 288	11 504 424	14·013
51	740 782	10 738 136	13·496
52	715 066	9 997 354	12·981
53	689 096	9 282 288	12·470
54	662 838	8 593 192	11·964
55	636 261	7 930 354	11·464
56	609 335	7 294 093	10·971
57	582 052	6 684 758	10·485
58	554 410	6 102 706	10·008
59	526 414	5 548 296	9·540

x	D_{xx}	N_{xx}	a_{xx}
60	498 096	5 021 882	9·082
61	469 492	4 523 786	8·635
62	440 669	4 054 294	8·200
63	411 701	3 613 625	7·777
64	382 694	3 201 924	7·367
65	353 758	2 819 230	6·969
66	325 035	2 465 472	6·585
67	296 677	2 140 437	6·215
68	268 855	1 843 760	5·858
69	241 744	1 574 905	5·515
70	215 534	1 333 161	5·185
71	190 403	1 117 627	4·870
72	166 535	927 224	4·568
73	144 096	760 689	4·279
74	123 231	616 593	4·004
75	104 067	493 362	3·741
76	86 690·9	389 295	3·491
77	71 157·7	302 605	3·253
78	57 483·9	231 447	3·026
79	45 643·3	173 963	2·811
80	35 571·2	128 320	2·607
81	27 167·7	92 748·4	2·414
82	20 300·4	65 580·7	2·231
83	14 812·8	45 280·3	2·057
84	10 533·8	30 467·5	1·892
85	7 283·99	19 933·7	1·737
86	4 885·25	12 649·7	1·589
87	3 169·01	7 764·49	1·450
88	1 982·08	4 595·48	1·319
89	1 191·06	2 613·40	1·194
90	684·867	1 422·34	1·077
91	375·099	737·469	·966
92	194·650	362·370	·862
93	95·114 8	167·720	·763
94	43·453 0	72·605 6	·671
95	18·403 3	29·152 6	·584
96	7·152 75	10·749 3	·503
97	2·520 48	3·596 57	·427
98	·793 328	1·076 09	·356
99	·218 978	·282 760	·291
100	·051 787	·063 782	·232

SECTION IV

PENSION FUND TABLES

The following tables are intended only for the use of students working examples in connexion with Pension Funds. The rates of decrement, the relative salary scale and the annuity values at retirement have been arbitrarily determined for this purpose only, and the tables cannot therefore be regarded as suitable for practical use in connexion with the valuation of or the calculation of rates of contribution for any actual fund.

PENSION FUND TABLES

INDEPENDENT RATES OF DECREMENT

Age x	Withdrawal	Death	Ill-health retirement	Age retirement	Age x
20	·030	·001 57			20
21	·027	·001 62			21
22	·024	·001 67			22
23	·021	·001 72			23
24	·018	·001 77			24
25	·015	·001 82			25
26	·012	·001 88			26
27	·009	·001 92			27
28	·006	·001 97			28
29	·003	·002 01			29
30		·002 07	·000 10		30
31		·002 12	·000 20		31
32		·002 17	·000 30		32
33		·002 22	·000 40		33
34		·002 27	·000 50		34
35		·002 33	·000 60		35
36		·002 40	·000 70		36
37		·002 49	·000 80		37
38		·002 60	·000 91		38
39		·002 73	·001 04		39
40		·002 88	·001 19		40
41		·003 06	·001 36		41
42		·003 27	·001 56		42
43		·003 52	·001 79		43
44		·003 80	·002 04		44
45		·004 13	·002 32		45
46		·004 49	·002 63		46
47		·004 88	·002 97		47
48		·005 30	·003 36		48
49		·005 75	·003 81		49
50		·006 24	·004 34		50
51		·006 78	·004 97		51
52		·007 39	·005 68		52
53		·008 10	·006 49		53
54		·008 90	·007 39		54
55		·009 80	·008 40		55
56		·010 80	·009 50		56
57		·011 90	·010 70		57
58		·013 10	·012 00		58
59		·014 40	·013 40		59
60		·015 90	·014 90	·200	60
61		·017 70	·016 50	·150	61
62		·019 90	·018 20	·100	62
63		·022 60	·020 00	·100	63
64		·026 00	·021 90	·100	64
65				1·000 (at 65 exact)	65

PENSION FUND TABLES

MULTIPLE DECREMENT TABLE

Age x	l_x	w_x	d_x	i_x	r_x	Age x
20	100 000	2 998	155			20
21	96 847	2 613	155			21
22	94 079	2 256	155			22
23	91 668	1 923	156			23
24	89 589	1 611	157			24
25	87 821	1 316	159			25
26	86 346	1 035	161			26
27	85 150	766	163			27
28	84 221	505	165			28
29	83 551	250	168			29
30	83 133		172	8		30
31	82 953		176	17		31
32	82 760		180	25		32
33	82 555		183	33		33
34	82 339		187	41		34
35	82 111		191	49		35
36	81 871		196	57		36
37	81 618		203	65		37
38	81 350		211	74		38
39	81 065		221	84		39
40	80 760		232	96		40
41	80 432		246	109		41
42	80 077		262	125		42
43	79 690		280	142		43
44	79 268		301	161		44
45	78 806		325	182		45
46	78 299		351	205		46
47	77 743		379	230		47
48	77 134		408	258		48
49	76 468		439	291		49
50	75 738		472	328		50
51	74 938		507	371		51
52	74 060		546	419		52
53	73 095		590	472		53
54	72 033		639	530		54
55	70 864		692	592		55
56	69 580		748	657		56
57	68 175		807	725		57
58	66 643		868	795		58
59	64 980		929	865		59
60	63 186		900	842	12 463	60
61	48 981		796	742	7 232	61
62	40 211		754	689	3 949	62
63	34 819		741	655	3 412	63
64	30 011		734	617	2 934	64
65	25 726				25 726	65

PENSION FUND TABLES

RELATIVE SALARY SCALE

Age x	s_x	$z_{x+\frac{1}{2}}$	Age x	s_x	$z_{x+\frac{1}{2}}$
20	1·000		45	3·000	2·854
21	1·104		46	3·054	2·913
22	1·206		47	3·106	2·970
23	1·306		48	3·156	3·025
24	1·404		49	3·204	3·078
25	1·500		50	3·250	3·129
26	1·594		51	3·294	3·178
27	1·686		52	3·336	3·225
28	1·776		53	3·376	3·270
29	1·864		54	3·414	3·313
30	1·950	1·729	55	3·450	3·354
31	2·034	1·818	56	3·484	3·393
32	2·116	1·905	57	3·516	3·430
33	2·196	1·990	58	3·546	3·465
34	2·274	2·073	59	3·574	3·498
35	2·350	2·154	60	3·600	3·529
36	2·424	2·233	61	3·624	3·558
37	2·496	2·310	62	3·646	3·585
38	2·566	2·385	63	3·666	3·610
39	2·634	2·458	64	3·684	3·633
40	2·700	2·529			
41	2·764	2·598			
42	2·826	2·665		$z_{65} = 3{\cdot}644$	
43	2·886	2·730			
44	2·944	2·793			

s_x is ratio of salary between ages x and $x+1$ to salary between ages 20 and 21.

$$z_{x+\frac{1}{2}} = \tfrac{1}{2}(z_x + z_{x+1}),$$

where $z_x = \frac{1}{5}(s_{x-5} + s_{x-4} + s_{x-3} + s_{x-2} + s_{x-1})$.

PENSION FUND TABLES

CONTRIBUTION FUNCTIONS **3** per cent

Age x	$D_x = v^x l_x$	$\bar{N}_x = \Sigma\frac{(D_x+D_{x+1})}{2}$	${}^s\bar{D}_x = s_x\frac{(D_x+D_{x+1})}{2}$	${}^s\bar{N}_x = \Sigma {}^s\bar{D}_x$	Age x
20	55 368	1 102 282	53 714	2 479 326	20
21	52 060	1 048 568	55 839	2 425 612	21
22	49 099	997 989	57 614	2 369 773	22
23	46 447	950 216	59 108	2 312 159	23
24	44 072	904 957	60 383	2 253 051	24
25	41 944	861 949	61 487	2 192 668	25
26	40 038	820 958	62 462	2 131 181	26
27	38 334	781 772	63 348	2 068 719	27
28	36 811	744 199	64 172	2 005 371	28
29	35 455	708 066	64 966	1 941 199	29
30	34 250	673 213	65 744	1 876 233	30
31	33 180	639 498	66 428	1 810 489	31
32	32 139	606 839	66 935	1 744 061	32
33	31 126	575 206	67 270	1 677 126	33
34	30 139	544 573	67 447	1 609 856	34
35	29 181	514 913	67 480	1 542 409	35
36	28 248	486 198	67 373	1 474 929	36
37	27 340	458 404	67 140	1 407 556	37
38	26 457	431 505	66 785	1 340 416	38
39	25 596	405 478	66 316	1 273 631	39
40	24 758	380 301	65 742	1 207 315	40
41	23 939	355 952	65 062	1 141 573	41
42	23 139	332 413	64 283	1 076 511	42
43	22 356	309 666	63 414	1 012 228	43
44	21 590	287 693	62 457	948 814	44
45	20 839	266 478	61 413	886 357	45
46	20 102	246 007	60 286	824 944	46
47	19 378	226 267	59 082	764 658	47
48	18 666	207 245	57 805	705 576	48
49	17 966	188 929	56 458	647 771	49
50	17 277	171 308	55 045	591 313	50
51	16 596	154 371	53 560	536 268	51
52	15 924	138 111	52 012	482 708	52
53	15 259	122 520	50 400	430 696	53
54	14 599	107 591	48 721	380 296	54
55	13 944	93 320	46 986	331 575	55
56	13 293	79 701	45 184	284 589	56
57	12 644	66 732	43 324	239 405	57
58	12 000	54 410	41 417	196 081	58
59	11 360	42 730	39 468	154 664	59
60	10 725	31 687	33 836	115 196	60
61	8 072	22 288	26 285	81 360	61
62	6 433	15 035	21 588	55 075	62
63	5 408	9 114	18 209	33 487	63
64	4 526	4 147	15 278	15 278	64
65	3 767				65

Age x	$\bar{a}^{i}_{x+\frac{1}{2}}$	$C^{ia}_x = v^{x+\frac{1}{2}} i_x \bar{a}^{i}_{x+\frac{1}{2}}$	$M^{ia}_x = \Sigma C^{ia}_x$	$\bar{R}^{ia}_x = \Sigma(M^{ia}_x - \frac{1}{2}C^{ia}_x)$	Age x
20			25 032	848 952	20
25			25 032	723 792	25
30	11·00	36	25 032	598 632	30
35	11·00	189	24 519	474 440	35
40	11·00	319	23 333	354 565	40
45	11·00	522	21 364	242 431	45
50	11·00	811	18 233	142 894	50
55	11·00	1 263	13 323	63 121	55
56	11·00	1 360	12 060	50 430	56
57	11·00	1 457	10 700	39 050	57
58	11·00	1 552	9 243	29 079	58
59	11·00	1 639	7 691	20 612	59
60	10·95	1 542	6 052	13 741	60
61	10·85	1 307	4 510	8 460	61
62	10·75	1 168	3 203	4 603	62
63	10·65	1 068	2 035	1 984	63
64	10·55	967	967	483	64

Age x	${}^{s}\bar{M}^{ia}_{x\rceil} = s_x(M^{ia}_x - \frac{1}{2}C^{ia}_x)$	${}^{s}\bar{R}^{ia}_x = \Sigma {}^{s}\bar{M}^{ia}_x$	${}^{z}C^{ia}_x = z_{x+\frac{1}{2}} C^{ia}_x$	${}^{z}M^{ia}_x = \Sigma {}^{z}C^{ia}_x$	${}^{z}\bar{R}^{ia}_x = \Sigma({}^{z}M^{ia}_x - \frac{1}{2}{}^{z}C^{ia}_x)$	Age x
20	25 032	2 036 249		80 823	2 814 406	20
25	37 548	1 885 556		80 823	2 410 291	25
30	48 777	1 674 786	62	80 823	2 006 176	30
35	57 399	1 412 327	407	79 819	1 603 886	35
40	62 567	1 113 526	807	77 064	1 210 937	40
45	63 309	797 099	1 490	71 795	837 484	45
50	57 938	488 454	2 538	62 472	499 846	50
55	43 784	222 892	4 236	46 606	223 855	55
56	39 648	179 108	4 614	42 370	179 367	56
57	35 058	139 460	4 998	37 756	139 304	57
58	30 024	104 402	5 378	32 758	104 047	58
59	24 557	74 378	5 733	27 380	73 978	59
60	19 012	49 821	5 442	21 647	49 465	60
61	13 978	30 809	4 650	16 205	30 539	61
62	9 549	16 831	4 187	11 555	16 659	62
63	5 503	7 282	3 855	7 368	7 198	63
64	1 779	1 779	3 513	3 513	1 757	64

All summations are based on the annual values (from age 64 or age 65 as the case may be) although for conciseness quinquennial values only are shown up to age 55.

PENSION FUND TABLES

AGE RETIREMENT FUNCTIONS **3 per cent**

Age x	$\bar{a}^r_{x+\frac{1}{2}}$ ($\bar{a}^r_x$ at 65)	$C^{ra}_x = v^{x+\frac{1}{2}} r_x \bar{a}^r_{x+\frac{1}{2}}$ ($=v^x r_x \bar{a}^r_x$ at 65)	$M^{ra}_x = \Sigma C^{ra}_x$	$\bar{R}^{ra}_x = \Sigma(M^{ra}_x - \frac{1}{2} C^{ra}_x)$	Age x
20			96 366	4 144 813	20
25			96 366	3 662 983	25
30			96 366	3 181 153	30
35			96 366	2 699 323	35
40			96 366	2 217 493	40
45			96 366	1 735 663	45
50			96 366	1 253 833	50
55			96 366	772 003	55
56			96 366	675 637	56
57			96 366	579 271	57
58			96 366	482 905	58
59			96 366	386 539	59
60	12·20	25 429	96 366	290 173	60
61	11·80	13 856	70 937	206 522	61
62	11·40	7 097	57 081	142 513	62
63	11·05	5 770	49 984	88 980	63
64	10·70	4 665	44 214	41 881	64
65	10·50	39 549	39 549		65

Age x	${}^s\bar{M}^{ra}_x = s_x(M^{ra}_x - \frac{1}{2}C^{ra}_x)$	$10^{-3}\,{}^s\bar{R}^{ra}_x = 10^{-3}\Sigma\,{}^s\bar{M}^{ra}_x$	${}^zC^{ra}_x = z_{x+\frac{1}{2}} C^{ra}_x$	${}^zM^{ra}_x = \Sigma^z C^{ra}_x$	$10^{-3}\,{}^z\bar{R}^{ra}_x = 10^{-3}\Sigma({}^zM^{ra}_x - \frac{1}{2}{}^zC^{ra}_x)$	Age x
20	96 366	10 823		346 377	14 907	20
25	144 549	10 243		346 377	13 175	25
30	187 914	9 431		346 377	11 443	30
35	226 460	8 413		346 377	9 711	35
40	260 188	7 211		346 377	7 980	40
45	289 098	5 850		346 377	6 248	45
50	313 189	4 355		346 377	4 516	50
55	332 463	2 748		346 377	2 784	55
56	335 739	2 416		346 377	2 438	56
57	338 823	2 080		346 377	2 091	57
58	341 714	1 741		346 377	1 745	58
59	344 412	1 400		346 377	1 399	59
60	301 144	1 055	89 739	346 377	1 052	60
61	231 969	754	49 300	256 638	751	61
62	195 181	522	25 443	207 338	519	62
63	172 665	327	20 830	181 895	324	63
64	154 290	154	16 948	161 065	153	64
65			144 117	144 117		65

All summations are based on the annual values (from age 64 or age 65 as the case may be) although for conciseness quinquennial values only are shown up to age 55.

3 per cent FUNCTIONS FOR PAYMENTS (WITHOUT INTEREST) ON DEATH

Age x	$C_x^d = v^{x+\frac{1}{2}} d_x$	$M_x^d = \Sigma C_x^d$	$\bar{R}_x^d = \Sigma(M_x^d - \frac{1}{2}C_x^d)$	${}^s\bar{M}_x^d = s_x(M_x^d - \frac{1}{2}C_x^d)$	${}^s\bar{R}_x^d = \Sigma {}^s\bar{M}_x^d$	Age x
20	85	4 224	108 130	4 181	246 988	20
25	75	3 823	88 035	5 677	222 874	25
30	70	3 461	69 838	6 681	192 296	30
35	67	3 118	53 398	7 250	157 598	35
40	70	2 780	38 647	7 411	120 856	40
45	85	2 405	25 657	7 089	84 219	45
50	106	1 939	14 755	6 129	50 426	50
55	134	1 357	6 462	4 451	22 826	55
56	141	1 223	5 172	4 017	18 375	56
57	147	1 082	4 019	3 548	14 358	57
58	154	935	3 010	3 042	10 810	58
59	160	781	2 152	2 505	7 768	59
60	151	621	1 451	1 962	5 263	60
61	129	470	906	1 468	3 301	61
62	119	341	501	1 025	1 833	62
63	113	222	220	605	808	63
64	109	109	55	203	203	64

All summations are based on the annual values (from age 64 or age 65 as the case may be) although for conciseness quinquennial values only are shown up to age 55.

3 per cent FUNCTIONS FOR PAYMENTS (WITHOUT INTEREST) ON WITHDRAWAL

Age x	$C_x^w = v^{x+\frac{1}{2}} w_x$	$M_x^w = \Sigma C_x^w$	$\bar{R}_x^w = \Sigma(M_x^w - \frac{1}{2}C_x^w)$	${}^s\bar{M}_x^w = s_x(M_x^w - \frac{1}{2}C_x^w)$	${}^s\bar{R}_x^w = \Sigma {}^s\bar{M}_x^w$	Age x
20	1 636	7 675	24 541	6 857	29 531	20
21	1 384	6 039	17 684	5 903	22 674	21
22	1 160	4 655	12 337	4 914	16 771	22
23	960	3 495	8 262	3 938	11 857	23
24	781	2 535	5 247	3 012	7 919	24
25	619	1 754	3 102	2 167	4 907	25
26	473	1 135	1 657	1 433	2 740	26
27	340	662	758	830	1 307	27
28	217	322	266	378	477	28
29	105	105	53	99	99	29
30 and over						30 and over

INDEX